SUNLIGHT INTO WINE

A Handbook for Winegrape Canopy Management

Richard Smart and Mike Robinson

under the auspices of
MINISTRY OF AGRICULTURE AND FISHERIES
NEW ZEALAND

Awarded
MENTION SPECIALE EN VITICULTURE
by the
OFFICE INTERNATIONAL DE LA VIGNE ET DU VIN
Paris 1992

Published by WINETITLES PTY LTD
630 Regency Road, Broadview, South Australia 5093
www.winebiz.com.au

First published in June 1991
Second printing September 1992
Third printing July 1994
Fourth printing March 1996
Fifth printing March 1997
Sixth printing March 1998
Seventh printing May 1999
Eighth printing September 2001
Ninth printing July 2003
Tenth printing February 2006
Eleventh printing April 2008
Twelfth printing September 2009

© Copyright 1991 Ministry of Agriculture and Fisheries, New Zealand

All rights reserved. No part of this publication may be copied or reproduced by any means with out the written permission of Winetitles Pty Ltd.

Front cover illustration: Sunlight streaming through an overhead vine canopy. Stellenbosch, South Africa. (Photo Richard Smart)

Back cover illustration: Mary Lee de Celles standing in a retrofitted U trellis vineyard, Napa Valley, California July 1989.

National Library of Australia Cataloguing-in-Publication

Smart, Richard E., 1945
Sunlight into Wine: a handbook for winegrape canopy management.

Bibliography.
ISBN 1 875130 10 1

1. Grapes-Growth. 2. Grapes-Climatic factors. 3. Viticulture. 4. Trellises.
5. Wine and wine making.
I. Robinson, M.D. (Michael D.) II. New Zealand
Ministry of Agriculture and Fisheries. III. Title.

634.8846

Wholly produced in Adelaide, South Australia
Designed by Michael Deves
Printed and bound by Hyde Park Press, Adelaide, Australia

This book is dedicated to wonderful friends in wine, and especially

PROFESSOR NELSON SHAULIS
of CORNELL UNIVERSITY

PIONEER OF CANOPY MANAGEMENT

Richard Smart (centre) and Professor Nelson Shaulis (right) discuss the finer points of canopy assessment.
Long Island, New York, July 1989.

TABLE OF CONTENTS

Authors' Preface

Over the last decade there has been an increased awareness of the role that canopy management can play in affecting winegrape yield and quality. This has been, and continues to be, an active area of research. As well, canopy management has become accepted as an important conference topic, a trend which began at the 1980 University of California Centennial Symposium. The concepts of canopy management have been particularly well promoted in the English-speaking New World.

Scientific studies have prompted commercial evaluations of new canopy management options for vinifera winegrape vineyards. Since the mid 1980s these evaluations have occurred in vineyards of California, Australia and New Zealand to name a few countries. We have invited some of these 'pioneers' from commercial viticulture to share their experiences with us in this handbook, to reinforce our message.

Further interest in canopy management has been generated by MAF New Zealand conducting canopy management workshops in different viticultural regions around the world. Until September 1990, there have been 20 of these held in New Zealand, Australia (four states), USA (six states), South Africa and England.

There are now many commercial viticulturists who wish to evaluate canopy management techniques. However, they are often not sure how to proceed. We have written this handbook especially to help these people. The format of this book was designed to be 'reader friendly'. Everywhere we try to avoid heavy scientific language, and the illustrations and plates should help to make concepts easily understood. For similar reasons we use both metric and imperial measures, apart from experimental data which are normally given in metric units only. We have also had to consider carefully the use of words which we know vary in their meaning from region to region. Thus chosen for use were the terms 'node' over 'bud' (the former is botanically more correct), 'cluster' over 'bunch' though we retain the phrase 'bunch rot', 'trellis system' over 'training system', 'flowering' over 'bloom' and 'trimming' over 'hedging'.

We resisted the temptation to include multitudinous references from the scientific literature in the text, in the interest of easy reading. Some key references are listed under 'Further Reading'.

The handbook is designed to help grapegrowers answer the following questions:

- Do I have a canopy problem?
- How will yield and quality be affected if I change my canopy management?
- Which is the best canopy management option to adopt?
- If I need a new trellis system, how can I construct it?

We firmly believe that it is possible to simply, cheaply and effectively assess winegrape canopies, and as a result develop management strategies to improve both yield and quality. Often people acknowledge that wine quality is created in the vineyard. However, at the moment only a few grapegrowers are prepared to take the most simple vineyard measurement as a form of quality assurance.

The techniques we present here will improve the quality and yield of many vineyards, and can be applied to both new and existing plantings. We hope they help you.

Richard E Smart

Richard Smart
Cassegrain Vineyards,
RMB 67, Pacific Highway,
Wauchope, NSW 2446, Australia.

Mike Robinson.

Mike Robinson
MAF Technology, Ruakura Agricultural Centre,
Private Bag,
Hamilton, New Zealand.

Richard Smart (Photo P.M.)

Mike Robinson (Photo G.H.)

New York State Agricultural Experiment Station
Cornell University,
Geneva, New York, U.S.A.

Foreword by Professor Nelson Shaulis

It is a pleasure to write a foreword to this handbook on grapevine canopy management.

The grapevine's canopy is a community of leaves. Canopy management is about many characteristics of that community influencing its size and density. Canopy size, as area or length per hectare, can affect grape yield. For a unit length of canopy, its density affects the sunlight and water environments of the leaves and grapes, and thus can affect the soundness and composition of grapes.

Canopy management is one of several 'managements' (as of soil, pests and crop) which should be integrated. This integration of canopy management is recognition that canopy characteristics can be affected by:

- pre-planting choices, as of variety, soil and site, row spacing, in-row spacing, row direction
- vine establishment and maintenance choices, as of training system, which locates the renewal area and the fruit zone, canopy height, vine size (as by crop load, soil management, nitrogen fertilisation, and irrigation) and pest control
- in season canopy-modifying choices, as shoot positioning, removal of surplus shoots, summer trimming and topping and leaf removal.

Our knowledge is incomplete, even of tradition-based canopy management, and the information transfer is inadequate, even of evidence we do have. Led by Dr Richard Smart, the Ruakura Agricultural Centre's grapevine canopy program combines research and teaching, as evidenced by this new handbook. Such efforts deserve our attention and encouragement, because they recognize that grape growers want to know more about canopies and grapes.

Nelson Shaulis
Professor Emeritus of Viticulture.

Professor Nelson Shaulis in a Concord vineyard belonging to S. Dudley, Fredonia, NY, September 1975. (Photo A.K.)
This photograph tells a story of canopy management. The vineyard is trained to the Geneva Double Curtain (GDC) system, with 2.7 m (9 ft) between rows and 2.4 m (8 ft) between the vines. Cordon height is 1.8 m (6 ft) and cordons are 1.2 m (4 ft) apart, giving 7290 m canopy per ha (9600 ft/ac). Cane prunings were 0.2 kg/m (0.13 lb/ft) canopy and yield 2.3 kg/m (1.54 lb/ft) canopy.

London, United Kingdom

Foreword by Jancis Robinson, MW

If any single person can be credited with forcing on the English-speaking wine world the long overdue realization that the vineyard is at least as important as the cellar — and in many ways more important — then that person is Richard Smart. His may not have been the first breakthrough in viticultural research but he has been one of the most energetic researchers and evangelists in the field, often literally. And above all, he has been able and keen to disseminate his knowledge and beliefs around the world.

In many ways it seems absurd that it has taken so long for canopy management to be valued as the sensitive winemaking tool that it is — and that there are still so many wine producers to whom the subject is no more than a name. As those in the Old World are becoming even more aware of wine quality from the New World, so New World producers are becoming more aware of Old World ideas on viticultural factors and their effect on wine quality. Although producers in the New World can plant the Old World's most famous varieties, they cannot hope to duplicate the precise combinations of soil and climate that have played such an important part in establishing the reputations of the most famous vineyards.

Dr Smart's ideas suggest, however, that even if exact duplication is impossible, the canopy microclimate of the famous vineyards can be copied most effectively. As a consumer who cares deeply about wine quality, I am thrilled by his and other experimental results which show how dramatically wine quality can improve by avoiding shade. Even more exciting is that leading New World producers are now using practices outlined in this book to commercial advantage.

On behalf of the wine consumers of the world I urge you to read on and try these approaches in your vineyards.

Jancis Robinson, MW

Jancis Robinson, MW

Acknowledgements

Publication of this booklet would not have been possible without the assistance of many people. The Ministry of Agriculture and Fisheries (MAF) New Zealand provided the facilities for much of the research reported here. We want to particularly acknowledge the technical contribution of past and present MAF staff — John Whittles, Joe Hoogenboom and Amirdah Segaran at Te Kauwhata Viticultural Station; Belinda Gould, Steve Smith, Joy Dick and Kate Gibbs at Ruakura Agricultural Centre; David Kitchen and Peter Wood at Manutuke Horticultural Research Station; and Terry McCarthy, Karen Snaddon and Larry Morgan at Hawke's Bay Agricultural Research Centre. The Department of Scientific and Industrial Research (DSIR) wine research team has also helped us understand wine quality implications of our field studies, and we want to acknowledge help from Reiner Eschenbruch, Tom van Dam, David Heatherbell, Wyn Leonard, Brent Fisher, Margaret Hogg and Kay McMath.

Initial drafts of this book were prepared while the senior author was employed at Ruakura. Subsequently, the work was completed under the auspices of Cassegrain Vineyards of Port Macquarie, NSW, and their contribution is gratefully acknowledged.

Many colleagues in viticultural science have contributed ideas and inspiration to this area of viticulture, in particular Dr Nelson Shaulis of Cornell University, Dr Mark Kliewer of the University of California, Dr Cesare Intrieri of the University of Bologna, and Dr Alain Carbonneau of INRA Bordeaux. Trellis construction is a discipline which has evolved from folklore, experience and engineering. Key contributors to this process in New Zealand have included Graham Pinnell of the New Zealand Agricultural Engineering Institute, Wiremakers Ltd, and many fencing contractors throughout New Zealand. Their collective contributions are acknowledged. Pauline Hunt of Ruakura Agricultural Centre has contributed much of the art work for figures, and Raegan Heta of Cassegrain Vineyards did the freehand artwork. Their skills are gratefully acknowledged. Most of the photographs were taken by Richard Smart, Barry Wylde of Ruakura and Mike Robinson. We also acknowledge photographs by Graeme Harrison, Amand Kasimatis, Kate Gibbs, Leopold van Huyssteen, David Jordan, Robert Emmett and Gary Wood. Maria Drake, Marion France, Kerry Argent and Sherri Cupitt typed the manuscript. David Jordan, Rich Thomas, Peter Dry, Mark Kliewer, Jim Wolpert, Charles McKinney, George Ray McEachern, Ron Blanks, Allan Clarke and Kate Gibbs made helpful suggestions on the manuscript. Eric Blackwell assisted with proof reading. Kate Gibbs also contributed ideas to several sections of the booklet, and further made many useful vineyard observations during the period she worked with Richard Smart. Clare Rafferty contributed early ideas to the layout, and Richard Smart and Michael Deves developed the final version. We are grateful to Brad Sayer and Rod East of MAF for arranging the book's production, and to Michael Deves of Winetitles for his professional input. We are grateful to all of these people, and to our families for their encouragement and tolerance during writing of this book.

We are pleased to acknowledge the contributions to this booklet from colleagues working in vineyards and wineries around the world. These include Eben Archer, Zack Berkowitz, Michael Black, Michael Brajkovich, Bob Campbell, John Cassegrain, Vanya Cullen, Joe and Mary Lee De Celles, Joy Dick, Phil Freese, Larry Fuller-Perine, Michelle Gandell, Rob Gibson, Kim Goldwater, Peter Gristina, David Jordan, Scott Henry, Diane Kenworthy, Colin Kidd, Mark Kliewer, David Lowe, Larry Morgan, Larry McKenna, Jancis Robinson, Nelson Shaulis, Steve Smith, Leopoldt van Huyssteen, Tony Wolf, Gary Wood, Peter Wood and Bruce Zoecklein.

About the Authors

Richard Smart studied Agricultural Science at Sydney University before taking up his first appointment in 1965 in viticultural research at Griffith, NSW. His initial research interests were in irrigation, vine physiology and canopy microclimate. Subsequently he completed an M.Sc. (Hons) degree at Macquarie University and a Ph.D. degree at Cornell University of New York under Professor Nelson Shaulis, studying canopy microclimate effects on yield. From 1975 to 1981 he taught at Roseworthy Agricultural College of South Australia. From 1982 to 1989 Dr Smart was responsible for the Government viticultural research program at the Ruakura Agricultural Centre at Hamilton, New Zealand. During this period he developed many of the ideas and techniques of canopy management presented here. He works as a Viticultural Scientist and Consultant, with clients in many countries.

Mike Robinson studied agricultural engineering at Lincoln College and has been employed by MAF New Zealand at the Ruakura Agricultural Centre as the rural structures consultancy specialist since 1979. He is a registered engineer and much of his work has involved research, development and consultancy focused on increasing the reliability and decreasing the cost of crop support and protection structures. The work was initially targeted at New Zealand kiwi fruit, but latterly has been in demand for many crops throughout Australia, the Pacific and the United States. His involvement has been at many levels: from writing manuals for fencing contractors to designs for individual grower clients, to development of proprietary systems for manufacturers and to lecturing on design rules and standards for engineers.

Winegrape Canopies and Their Importance

'Sunlight into Wine' — an Explanation

'A day without wine is like a day without sunshine' . . . so says the old proverb. This handbook discusses specifically the relationship between sunlight and wine.

Wine is a product of sunlight. Grapevine leaves use energy from sunlight to change carbon dioxide (CO_2), an atmospheric gas, into sugars. This process is called photosynthesis. From the leaves the sugars move to the fruit. At a desired ripeness, grapes are harvested and crushed at the winery to produce juice as a first step in winemaking. Yeast cells convert sugars in the juice into alcohol during the process of fermentation and the juice is transformed into wine. And so, the close association between sunlight and wine can be seen. Warm and sunny climates cause grape juice to have a high sugar content and so produce wines with higher alcohol.

This handbook is, however, concerned with indirect relationships between sunlight and wine — that is, the effect of sun exposure of grape clusters (bunches) and leaves on wine quality. To explain this relationship, we should first think of some Old World ideas about factors that affect wine quality. From these regions we often hear phrases like 'a struggling vine makes the best wine', or 'low grape yield gives high wine quality'. A common feature of both 'struggling' and 'low-yielding' vines (the two are often synonymous) is that they grow few leaves. The leaves that do grow are smaller than for non-struggling or high-yielding vines. These vine canopies are then less shaded, and most leaves and fruit are well exposed to the sun. This observation leads to the important question — does a vine need to be struggling or low-yielding to make high quality wine, or, perhaps, is it necessary for the leaves and clusters to be well exposed to the sun?

This handbook is based on a **yes** answer to the second part of the question. Research in New Zealand and, indeed, in many countries has shown that dense, shaded canopies reduce wine quality. Changing the canopy so that clusters and leaves are better exposed to the sun has been shown to improve wine quality and yield.

This result contradicts the firmly held traditional beliefs listed earlier. We make no apologies for this. There is clear scientific and commercial experience to support our viewpoint.

In the pages to follow, we expand on these ideas. First, we discuss how grapevines function, and respond to their climate. Then the concept of canopy microclimate is discussed, to illustrate the factors that are important in leaf and cluster exposure.

Following this we show how the winegrape grower can assess his or her own vineyards to answer the question 'Do I have a canopy problem?'.

Then finally, some practical canopy management solutions to the problem, are presented. Generally the answer involves a change in trellis method. Details of the construction of the most important trellises are provided, along with tips to carry out the task in a cost efficient manner.

Readers will be able to use ideas presented here to increase yield and to make improved quality wine by a process we call **'winemaking in the vineyard'**.

Sunlight streaming through an overhead vine canopy. Stellenbosch, South Africa. (Photo R.S.)

Searching for a cluster of grapes in a dense canopy! Fresno, California. (Photo R.S.)

The interior of a shaded winegrape canopy, South Australia. (Photo R.S.)

Practical session during canopy management workshop, Long Island, N.Y., July 1989. (Photo R.S.)

Canopy Management — What is it?

What is a grapevine canopy?
A grapevine canopy is the above ground part of the vine formed by the shoot system. It includes shoots (leaves, petioles, shoot stems, shoot tips, lateral shoots and tendrils) and the fruit, trunk and cordon or canes. In vineyards, canopies are **continuous** when the foliage from adjacent vines down the row intermingles. When individual vines have canopies separated, they are termed **discontinuous**. Canopies are **divided** when the canopies of one vine — or even adjacent vines — are divided into discrete curtains or walls of foliage.

Dense or **crowded** canopies have excess leaf area in the canopy and so are shaded. **Open** or **low** density canopies are not shaded. All canopies have an outer or **exterior** layer of leaves. Not all canopies have **interior** leaves, ie. those between the two surfaces of exterior leaves. A **dense** canopy has a high proportion of interior leaves.

What is canopy management?
Canopy management includes a range of techniques which alter the position and number of shoots and fruit in space. In other words, canopy management is manipulation of canopy microclimate as will be subsequently defined. As well, canopy management can aim to alter the balance between shoot and fruit growth.

Techniques of canopy management include:
- **winter pruning which affects future shoot location and** density,
- **shoot thinning or desuckering** which affects shoot density,
- **summer pruning** (trimming) which shortens shoot length,
- **shoot devigoration** which aims to reduce shoot length and leaf area,
- **shoot positioning** which determines where shoots are located,
- **leaf removal** which is normally done around the cluster zone, and,
- **trellis system** changes which typically are designed to increase canopy surface area and to reduce canopy density.

What are the benefits of canopy management?
The following sections introduce the benefits of canopy management. These can be summarized as:
- improving wine quality,
- improving winegrape yield,
- reducing the incidence of some diseases, and
- decreasing production costs, especially by facilitating mechanization.

How are canopy management techniques applied?
Winegrape growers need to decide when and where application of canopy management principles are required. Typically these decisions will be made on a block by block basis within the vineyard.

The following sections will show you:
- how to identify problem canopies,
- how to decide which canopy management options to use, and
- how to successfully implement this option.

Throughout we emphasis practical management aspects.

The Rothbury Estate
Pokolbin, NSW, Australia

Retrofitted RT2T on Sauvignon Blanc vines at Cowra Vineyards. Photographed before shoot positioning. Note previous posts in line with vine trunks. (Photo R.S.)

Geneva Double Curtain retrofitting as used by Rothbury Estate. Photograph in Nigel Read's vineyard, Hawkes Bay. (Photo R.S.)

David Lowe

Rhine Valley vineyards facing south. An example of mesoclimate importance. Rudesheim, Germany. (Photo R.S.)

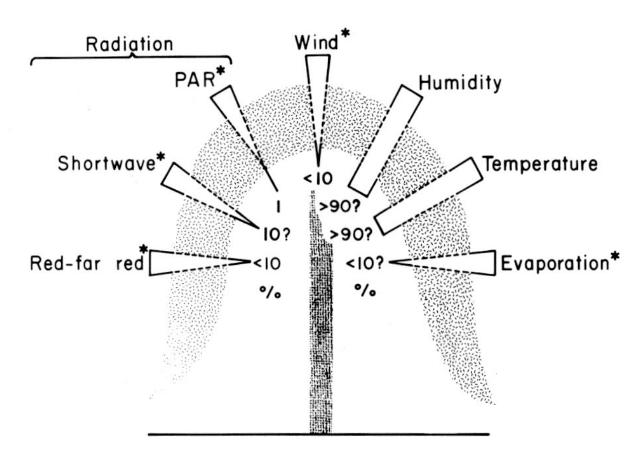

Relative differences in climate levels from the outside to the inside of dense grapevine canopies. From Smart (1984).

Measuring the red to far red ratio of sunlight in a shaded canopy. Rukuhia, New Zealand. (Photo R.S.)

Grapevine Canopy Microclimate

Three levels of climate
Early observers of winegrapes noted a strong effect of climate on both yield and quality. To fully understand these effects we need to distinguish between three levels of climate based on the ideas of Geiger, the German climatologist.

Macroclimate (or regional climate)
This is the climate of a region, and the general pattern is described by a central recording station. The macroclimate definition can extend over large or small areas, but typically the scale is tens of kilometres depending on topography and other geographic factors, e.g. distance from lakes or ocean.

Mesoclimate (or topo or site-climate)
The mesoclimate of a particular vineyard will vary from the macroclimate of the region because of differences in elevation, slope, aspect or distance from moderating factors like lakes or oceans. Mesoclimate effects can be very important for the success of the vineyard, especially where climatic conditions are limiting. For example, in the Mosel Valley (West Germany), a region marginal to ripen grapes, vineyards are planted on hillsides facing south to promote sunlight absorption. Another example is in New York State where the vineyards are located on lake shores to reduce winter freeze injury. Vineyards planted at high elevations can lead to cooler conditions for improved table wine quality, compared to the hotter valley floor. Examples of this effect are found in the Barossa Ranges of South Australia and Napa and Sonoma Valleys in California. Differences in mesoclimate can occur over ten to hundreds of metres, or by up to several kilometres.

Microclimate (or canopy climate)
Microclimate is the climate within and immediately surrounding a plant canopy. Measurements of climate show differences between the within canopy values, and those immediately above it, (the so called 'ambient' values). A simple example is sunlight. Sunlight can be measured at the centre of a dense canopy as one percent or less of the values measured above the canopy. Microclimate differences can occur over a few centimetres.

There is confusion in the popular literature (and on wine bottle back labels!) about the term microclimate. Often the word microclimate is used incorrectly when in fact mesoclimate would be more accurate. Thus, it is better to say 'this vineyard has a special **mesoclimate** due to its aspect', rather than 'a special **microclimate**'. When we consider the climate of a vineyard we are concerned with the macroclimate or mesoclimate. When we consider the climate of or within an individual vine, or part of a vine like a grape cluster, microclimate is appropriate.

Why do canopy microclimates differ?
The canopy microclimate is essentially dependent on how dense (or crowded) is the canopy. Let us now consider each climate element in turn, to show how they are affected by canopy density.

Sunlight quantity
The amount of sunlight falling on a vineyard varies with latitude, season, time of day and cloud cover. Sunlight intensity is commonly measured in units that correspond to the ability of plants to use sunlight in photosynthesis. Consequently, the intensity is often termed 'photosynthetically active radiation' (or PAR). The units are amounts of energy per unit area per unit time, i.e. micro Einsteins per square metre per second, $\mu E\ m^{-2}s^{-1}$. A bright sunny day might give readings over 2000 $\mu E\ m^{-2}s^{-1}$, and overcast conditions can reduce this value to less than 300 $\mu E\ m^{-2}s^{-1}$.

Values of sunlight intensity measured in the centre of dense canopies can be less than 10 μE m^{-2}s^{-1} although above canopy (ambient) values are over 2000 μE m^{-2}s^{-1}. The reason for this large reduction is that grapevine leaves strongly absorb sunlight. Measurements show that a leaf in bright sunlight (say 2000 μE m^{-2}s^{-1}) will only transmit 6%, so 120 μE m^{-2}s^{-1} pass through to the next leaf layer in the canopy. A third leaf in line would receive only 7 μE m^{-2}s^{-1} and would be in deep shade. This simple example ignores reflection of light between leaf layers.

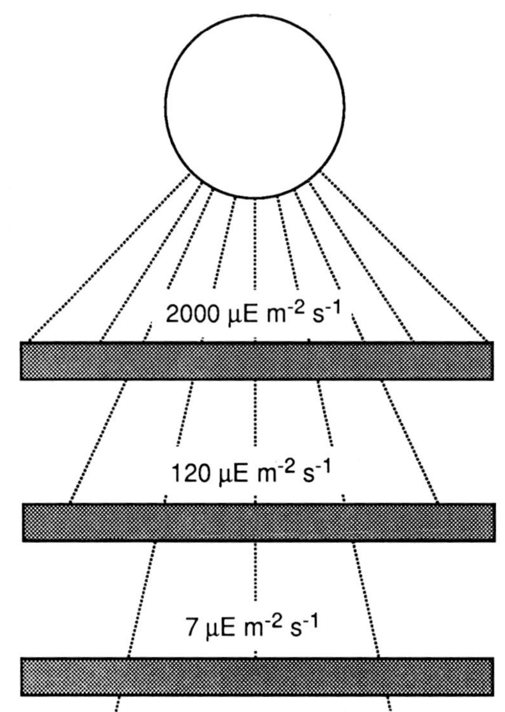

Reduction of sunlight by transmission through two grapevine leaves.

Sunlight quality

Not only is the amount of sunlight altered in the canopy, but so is the colour spectrum that makes up sunlight. Plant leaves absorb only a part of the sunlight—especially in the so called 'visible range' (400-700 nm wavelength). This is the part of the sunlight spectrum we can see. So, as sunlight passes through the canopy, there is less sunlight in the visible range relative to the remaining wavelengths of the spectrum. An important consequence is that the ratio of red light (660 nm) to far red light (730 nm) declines in the canopy. Plants like grapevines respond to the red:far red ratio via their **phytochrome** system, and this is important in affecting, for example, fruit colour development.

Although shoots, stems, petioles and fruit also absorb sunlight, by far the greatest reason for shade in grapevine canopies is the sunlight absorbed by leaves.

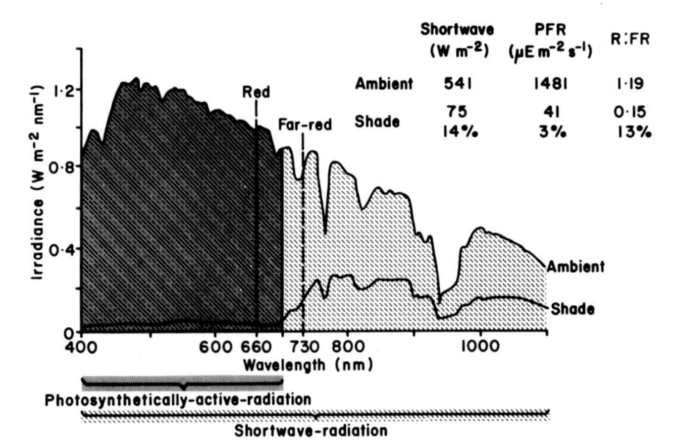

The spectral distribution of sunlight above ('ambient') and below ('shade') a dense grapevine canopy. Note the division into photosynthetically active-radiation (PAR) and shortwave radiation, and the red and far red wavelengths. Te Kauwhata, New Zealand.

Temperature

Temperatures of grapevine parts are generally at or near air temperature. This applies unless they are warmed by absorbed sunlight, or cooled by evaporation of water, as takes place with transpiration from leaves.

Elevated tissue temperatures due to warming by sunlight are most obvious on sunny and calm days. For example, grape berries exposed to bright sunlight on calm days can be warmed up to 15°C (27° F) above the air temperature. Wind cools because it removes some of the absorbed heat from the leaf.

In contrast, grapevine leaves do not warm as much as berries because leaves are cooled by transpiration, whereas berries have little transpiration to show this effect. Transpiration is a process whereby water is vaporized within the leaf surface and escapes through stomata (pores) on the bottom of the leaf. The vaporization or change from liquid to gas requires energy (heat) which cools the leaf in the same way that an evaporative air conditioner works. As long as the vines are well supplied with water, leaves exposed to full sunlight are typically less than 5°C (9° F) above air temperature. If the vine is water-stressed, the temperatures can be higher. It is interesting to note that leaves on the shaded side of canopies can even be below air temperature. These leaves still transpire but do not have the heat absorbed from the sun.

At night, the outside parts of the canopy can lose heat to the atmosphere, especially on clear calm nights. This is known as **longwave cooling**. Under these conditions, exterior leaves and berries can be cooled to 1-3°C (2-5°F) below air temperatures.

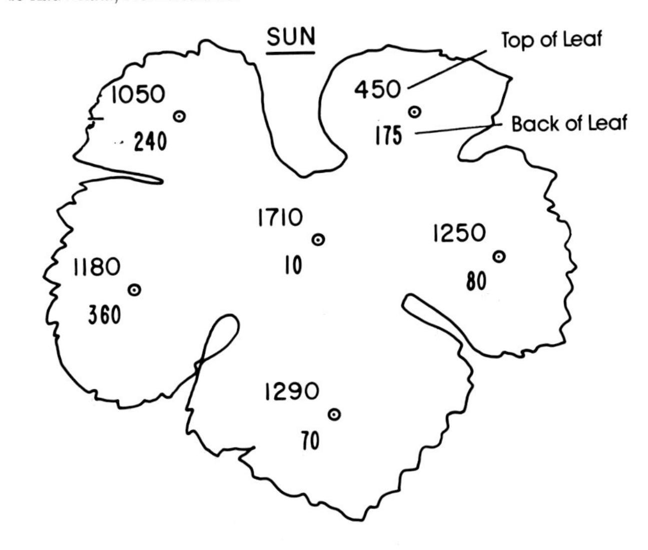

Spot measurements of sunlight (PAR) on the top and back of a grapevine leaf. Top values vary due to differences in sun-leaf angle. Units are μEm^{-2}s^{-1}. Roseworthy, South Australia. (Smart, unpublished data).

Humidity

Transpiration by leaves can lead to a slight build up of humidity inside dense canopies. If the canopy is open, then the ventilation effect of even a slight breeze can reduce the humidity difference from the canopy exterior to interior. However, even small differences in humidity can be important for the establishment of fungal diseases like **Botrytis**.

Wind speed

Wind patterns around a vineyard are complex, and there is an interaction between wind direction and row orientation. Similar to sunlight, wind speed is very low in the centre of dense canopies. This occurs because leaves slow down air flow. Measurements in New Zealand show that wind speed in the centre of dense Semillon canopies can be less that 10% of the above canopy value.

Evaporation
Evaporation of free moisture (e.g. dew, rain) from plant surfaces is encouraged by high values of temperature, sunlight and wind speed, and low relative humidity. Hence, there can be significant differences in evaporation rates between the canopy exterior and interior. This is very important for fungal disease development, as the rate at which dew or rainfall evaporates will again depend on canopy density.

Microclimate and Canopy Density

	Open canopy	Dense canopy
Sunlight	Most leaves and fruit exposed	Most leaves and fruit in shade
Temperature	Fruit and leaves can be warmed by sunlight. At night, outside leaves and fruit can be cooled.	Most leaves and fruit are interior so are close to air temperature day and night.
Humidity	Leaves and fruit experience ambient humidity values.	Humidity can slightly build up in the canopy.
Wind speed	Leaves and fruit exposed to about ambient values	Wind speeds reduced in the canopy.
Evaporation	Evaporation rates similar to ambient values	Evaporation rates are reduced in the canopy

PAR sensor, exposed to full sunlight. Reading 1100 $\mu Em^{-2}s^{-1}$, Rukuhia, New Zealand. (Photo B.W.)

Cover the sensor with a Cabernet Franc leaf to measure transmission only. PAR reading 60$\mu Em^{-2}s^{-1}$, transmission = 5.5%. (Photo B.W.)

Red to far red sensor exposed to full sunlight. Reading 1.15. (Photo B.W.)

Cover the sensor with a Cabernet Franc leaf to show altered red to far red ratio of transmitted light. Reading 0.17. (Photo B.W.)

Canopy Microclimate and Grapevine Physiology

This section describes the interaction of canopy microclimate and grapevine physiology. It will become obvious that sunlight has a major influence on the way grapevines function.

Transpiration

Water absorbed by the roots is drawn into the leaves from where it evaporates in a process known as **transpiration**. On the under side of grapevine leaves are microscopic pores called **stomata**. These pores open during the day, and close at night, thus controlling transpiration. The rate of transpiration is closely linked to the climate, being highest under sunny, hot, windy and low humidity conditions.

Stomata open and close in response to sunlight. They begin to open with very low light levels soon after dawn and are fully open at PAR of about 200 μE $m^{-2}s^{-1}$. As temperatures and sunlight increase and humidity falls, transpiration increases. The stomata may partially or totally close when the vine cannot supply sufficient water to meet the transpirational demand. This closure most often occurs in the early afternoon. As light levels decrease just before dusk, stomata close again and remain closed during the night.

Exterior leaves on a grapevine canopy are exposed to higher sunlight levels and temperature, and thus transpire more than shaded interior leaves. Interior leaves in dense canopies may be exposed to such low light levels that their stomata do not completely open.

Photosynthesis

Photosynthesis is the process by which the energy from the sun is used by the green tissue of plants to convert carbon dioxide (CO_2) an atmospheric gas, to sugars. These sugars are the basic building blocks of most chemical materials found in the grapevine. These materials include carbohydrates, proteins, phenols, organic acids plus many others. Photosynthesis occurs mostly in the leaves and CO_2 diffuses into the leaf cells mainly through the stomata.

Photosynthesis is dependent on sunlight. In grapevines, no photosynthesis occurs at low light levels, below about 30 μE $m^{-2}s^{-1}$, or about 1½ % of full sunlight. Part of this reduction is due to stomata being partly closed reducing inflow of CO_2. As light intensity increases, so does photosynthesis, until about one third full sunlight is reached, or 700 μE $m^{-2}s^{-1}$. At this intensity photosynthesis in grapevine leaves is termed **light saturated**, and remains at about the same rate with even higher sunlight levels. The rate of photosynthesis also depends on leaf temperature, with maximum rates between 20 to 30° C (68 to 86° F). Photosynthesis is inhibited by low temperatures, less than 10° C (50° F) and by high temperatures, greater than 35° C (95° F).

Due to low sunlight intensity, interior leaves have low photosynthetic rates, which means they contribute little to the vine. When in deep shade, interior leaves turn yellow and become incapable of photosynthesis. Studies have shown that the exterior leaves contribute most photosynthesis in dense canopies. Green grape berries can also photosynthesize but their contribution is much smaller than that from leaves.

Joy Dick measuring grapevine photosynthesis. Rukuhia, New Zealand. (Photo B.W.).

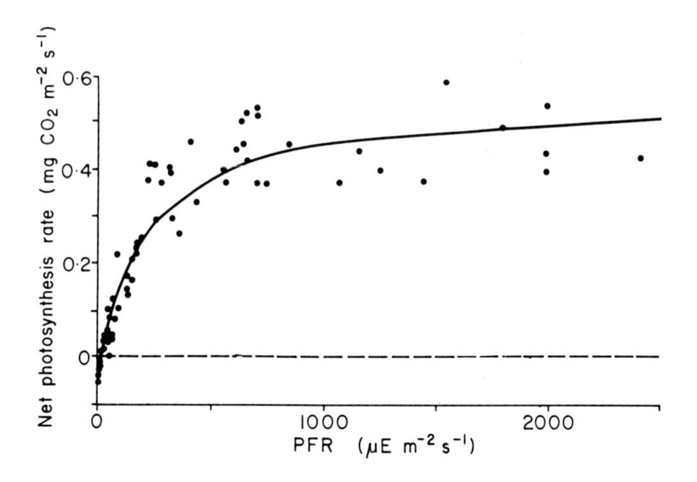

The relationship between sunlight measured as PAR and grapevine photosynthesis. From Smart (1985).

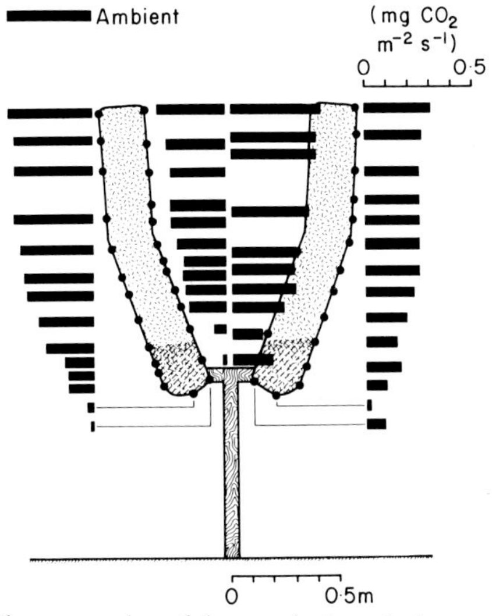

Estimated daily average values of photosynthetic rate for leaves at the canopy exterior based on PAR measurements. Note that leaves at the bottom of the canopy show low rates due to shading. From Smart (1985).

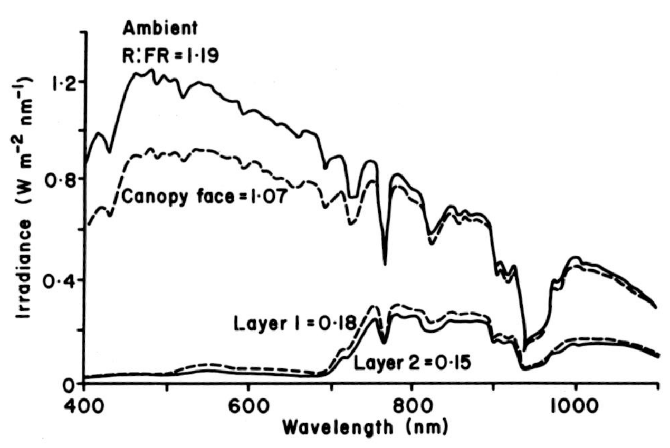

Spectral distribution of sunlight measured above, at the canopy face and at leaf layers one and two. Note the rapid loss of PAR and change of red to far red ratio. (From Smart 1987)

Michelle Gandell measuring transpiration rates of grapevine leaves. Rukuhia, New Zealand. (Photo B.W.).

Canopy Effects on Grapevine Physiology

	Open canopy, or exterior of dense canopy	Interior of dense canopy
Transpiration	Higher rate of transpiration.	Lower rate of transpiration.
Photosynthesis	Higher rate of photosynthesis.	Lower rate of photosynthesis.
Respiration	Often high rate of respiration during the day due to sunlight warming	Daytime respiration rate is largely controlled by air temperature.
Translocation	Exterior leaves are exporters of photosynthesis products.	Interior leaves are importers of photosynthesis products.
Phytochrome	Fruit and leaves are exposed to higher ratios of red to far red light.	Fruit and leaves are exposed to lower ratios of red to far red light.
Water relations	Exposed leaves and fruit experience more water stress.	Interior leaves and fruit experience less water stress.

Respiration

Plants require an internal source of energy to grow and manufacture complex chemical molecules. This chemical energy is produced by **respiration** where sugars and other compounds and oxygen interact and produce energy along with water and CO_2. Hence, this process is the reverse of photosynthesis. Of particular interest to winemakers is the respiration of malic acid as grape berries ripen. Consequently, the concentration of malic acid declines with time.

Respiration is very temperature dependent. Normally for each 10°C (18°F) increase in temperature, the respiration rate doubles. Hence malic acid levels in berries are lower in warm than cool viticultural regions.

Translocation

Translocation is the process by which chemical materials and nutrients are moved in the vine. For example, sugars manufactured by photosynthesis in leaves are translocated to other parts of the vine. Sugar can be exported to either the shoot tip, the grape cluster, the root system and/or other permanent parts like the trunk for storage. Translocation rates are not as sensitive to microclimatic conditions as are some other processes, but shaded shoots are known to import sugars to provide energy for growth.

Phytochrome

Plants use **phytochrome** reactions to alter their growth habit in response to the light environment. A classic response is the way a grape shoot will grow long, thin internodes in the shade, in an attempt to reach the light. Phytochromes are hormone-like substances within the vine. Their levels are determined by the ratio of red to far red wavelengths of light in the environment. Thus, leaves and fruit in the centre of dense canopies receive low levels of red to far red light (about 0.1 to 1). This is very different to the ratio at the canopy exterior of about 1.2 to 1.

Phytochrome effects are not as well understood as are those of photosynthesis. However, recent studies in New Zealand showed that phytochrome affects red colour (anthocyanin) formation and sugar levels in Cabernet Sauvignon berries. It is possible that some effects of shade on fruit composition may be due to light quality as well as quantity. Perhaps grapes are like other plants where it is known that phytochrome affects enzymes which govern levels of glucose, fructose, phenolics, malic acid, K and hence pH. More research on this subject is required.

Water relations

Whether a grapevine is water stressed or not depends on the water supply from the root zone, and the evaporative demand of the atmosphere. Water stress occurs when soil water supply is low, and when the weather is hot, sunny, windy and of low humidity.

Exterior leaves of the canopy experience more water stress than interior leaves. This is because they are exposed to more sunlight and higher wind speeds and so transpire more than interior leaves. Consequently a large, open, well exposed canopy will use more water than a restricted, dense and shaded one, and shows more water stress symptoms.

Altering the trellis to increase canopy surface area leads to more vineyard water use. Unless matched by increased irrigation or soil water supply there will be more water stress.

The Growth of Grapevines

Annual growth cycle

The grapevine is a perennial plant and, when well tended, can live for a very long time. The famous vine in Hampton Court, England was planted in 1769 and still grows. Most commercial vineyards, however, are replanted within 50 years.

Each spring, grapevines begin growth with **budburst**. This usually occurs when mean daily temperatures are about 10°C (50°F). At first shoots grow slowly, but as temperatures increase they elongate more rapidly. This is the so-called 'grand' period of growth. Early shoot growth is dependent on stored reserves in the vine. As shoots elongate and leaves mature then current photosynthesis provides for further shoot and fruit growth. Location and variety determine when **flowering** (bloom) occurs. Usually it is 30 to 80 days after budburst when the average temperature is about 16 to 20°C (61 to 68°F).

Berry development begins with **berry set** and ends with **harvest**, this period lasting 70 to 140 days, again dependent on location and variety. **Veraison** is the stage mid-way through berry development when berries change from being green and hard to coloured (yellow, pink, red or black) and soft. At this transition sugars begin to accumulate in the berry. In hot climates the total period from budburst to harvest is short (110 to 140 days), but is much longer for cooler climates (190 to 220 days). Leaf fall is stimulated by frost or water stress. After leaf fall the vine is dormant over winter.

The grapevine root system also has pronounced growth cycles. Most fruit trees have a flush of growth before budburst. However, for grapevines a first flush occurs at about flowering and there is a second flush at about harvest.

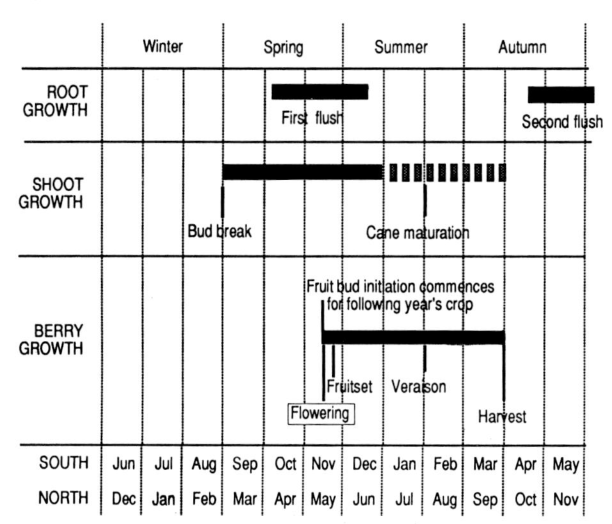

Patterns of root, shoot and berry growth for the grapevine, for northern and southern hemispheres.

Shoot development

An understanding of leaf production on shoots is important because leaves are largely responsible for creation of shade in grapevine canopies. Once grapevine leaves grow to about half their final size and lose their initial bright green colour, they begin to export sugars generated from photosynthesis. Before being net exporters, they rely on the import of sugars from stored reserves or photosynthesis by other leaves to supplement their own photosynthesis for growth.

A shoot has only one growing point, when it commences growth, the shoot **apex**. Young leaves unfold from this apex as the shoot grows. Often, as the fruit ripens the shoot tip will slow down or stop growing. However, under conditions of high vine vigour, and with vineyards on soils well supplied with water and nitrogen, shoot growth may continue until after harvest.

A lateral shoot is produced at each leaf node along the shoot. Each developing shoot therefore has many potential growing points, comprising the shoot apex and lateral shoots. These laterals may or may not develop any further than a few millimetres long. However, in many situations of high vine vigour, laterals grow vigorously and can develop to several metres in length.

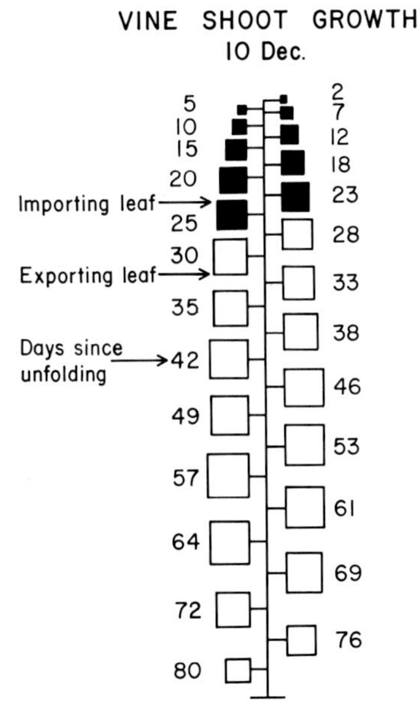

Schematic diagram of a grapevine shoot showing relative leaf areas, whether the leaves import or export products of photosynthesis, and their age. Variety Shiraz, Griffith, NSW, sampled 10 December. (Smart, unpublished data).

Shoot vigour

A common problem in many vineyards is excessive shoot vigour. Shoots with high vigour grow rapidly and have large leaves, long internodes, are thick (large diameter) and have active lateral growth at many nodes. When the shoot tip is removed by trimming, growth of the laterals near the cut shoot end is stimulated. A high vigour shoot can grow to more than 4 m (12 ft) length if not trimmed. A low vigour shoot will stop growth early in the season, and may be as short as 150 mm (6 in). For unirrigated vineyards in Mediterranean climates, shoot growth often stops soon after flowering when shoots are about 1 m (3 ft) long.

Shoot growth is encouraged by training upwards and is retarded by training downwards. Thus, one way to devigorate shoots is to train them downwards, as occurs with the Geneva Double Curtain and Scott Henry trellis systems (see later sections). Vigorous shoot growth is also encouraged when vines are pruned in winter to too few nodes (buds). Soils well supplied with water and nitrogen throughout the growing season cause vigorous shoot growth.

A rapidly growing shoot tip. (Photo R.S.)

Shallow rooting depth due to shallow ploughing and natural subsoil compaction. When this soil was deep ploughed, it became a high potential soil for vineyards. South Africa. (Photo L.H.)

A typical example of a well distributed root system on a calcareous sandy loam soil after effective soil profile modification. South Africa. (Photo L.H.)

Leopoldt van Huyssteen.

Growth balance in grapevines

There is a well known link between root growth and shoot growth. A vine which has a large, healthy root system will produce an extensive shoot system. Such a vine if pruned too severely in winter, say to about 40 nodes, will grow very vigorous shoots and also many water shoots will develop (shoots that develop from older buds on the cordon or head). Typically, a crowded canopy results. Poor soil conditions or the presence of pathogens such as phylloxera or nematodes restrict root growth. Consequently, with a small root system these vines have reduced shoot growth.

There is also a balance between shoot growth and fruit growth. For example, if clusters have a poor set of berries that limits the amount of fruit, then shoot vigour is encouraged. Conversely, pruning to many nodes in winter produces large crops and spindly shoot growth.

Viticultural and Oenological Research Institute
Stellenbosch, South Africa

Soil Management as a Tool for Managing Vine Vigour

Grapevines are adapted to a wide range of soil types, but like all plants, they grow best when they have healthy and well-developed root systems. The optimum rooting volume must supply sufficient water, be well aerated, have a low mechanical resistance, have optimum temperatures and pH, and contain no toxic substances. Many vineyard soils in their natural state have a low **potential** for the growing of grapevines due to one or more of the above properties not being optimal, especially in the subsoil. Vines then develop shallow root systems and are prone to drought. When such soils are undermanaged, overall poor vine performance and areas of uneven growth are evident. Under such conditions there is little scope for canopy management.

Soil properties affecting root growth should be properly assessed in site selection for vineyards, and should determine the appropriate soil management (SM) techniques. With SM, such as profile modification by deep soil preparation and soil tillage for weed control and water conservation, we can alter the soil to better integrate climatic effects, for example, water storage, drainage and soil temperature. In doing so the soil potential for sustaining vegetative growth is altered.

It was established in several South African SM experiments that vine vigour increases linearly with increasing soil depth due to larger rooting volumes. Maximum vigour may be unwanted due to canopy management problems, and there is no use in expanding the grapevine root system beyond that necessary to provide the water and nutrients to permit optimal utilization of the climate potential, i.e. sunlight. Without experience, it is almost impossible to predict beforehand what soil depth will result in the optimum vigour for a particular site and soil type. It is therefore recommended that any vineyard soil with subsoil problems should be loosened to a depth of at least 800 mm to homogenize the soil. Should the vine vigour be too strong to handle with viticultural practices such as plant spacing and trellising, further SM can be employed to reduce vigour, for instance, by regulating the competition effect of cover crops. This can be done by using different widths of cover crop strips or through killing the cover crop later in the growing season of the vine. In cases where there is not enough vigour due to too shallow soil preparation, loosening and thus root pruning in the middle of every second row should be done. A cover crop mulch should then be established on the surface to conserve soil water, and to keep soil temperatures in the topsoil favourable for root growth.

Although there is no universal system that will suit every vineyard, an understanding and application of the available SM options will allow a wide spectrum for controlling vine vigour.

L. van Huyssteen
Dr Leopoldt van Huyssteen
Soil Scientist

Canopy Microclimate and Yield

Grapevine yield is a result of a series of processes that take place over a period of about 17 months before the grapes are harvested. All of these processes are affected by shade in the canopy even though they may operate independently of each other.

Cluster initiation

The buds that contain the shoots and clusters for the following season begin growth early in the spring as shoots develop. Flowering is a critical stage. It is at this stage that the cluster **primordia** are laid down. Cluster primordia are microscopic pieces of tissue which eventually develop into clusters the following season. Within the young 'compound' bud there are three small shoots which develop—the primary, secondary and tertiary. Whether or not these shoots produce cluster primordia is very dependent on shade in the canopy. Studies in Australia with Sultana have suggested that light on the node itself is important in triggering this response. However, studies with the American variety, Concord, in New York State suggest that the leaf beside the node receives the light stimulus. Whichever occurs, it is still important that canopy shade be minimized on the sections of shoots that will be pruned to in the winter, as the light response is very localized. The basal part of shoots is called the **renewal** zone.

Budburst

Budburst largely depends on the number of nodes left at winter pruning. We use the term **node** rather than **bud** since in fact there can be up to three buds under a common scale at each node. Percentage budburst is defined as the number of shoots (multiplied by 100) divided by the number of nodes left. Typically, when many nodes are retained there is a smaller percentage budburst than when few nodes are retained. When vines with a large root system (i.e. high capacity) are pruned to a few nodes then, typically, many **water shoots** are produced and percentage budbreak is often greater than 100. Water shoots arise out of old buds left on the permanent wood of the vine (trunk or cordons). For many varieties these shoots have no or few clusters. The proportion of water shoots decreases when the node number retained at winter pruning is increased.

Recent studies have identified that budbreak is reduced in shaded canopies. Presumably this effect relates to shade on the shoot in the previous season. It is common to find buds that do not burst in the centre of dense canopies.

Flowering and fruit set

Fruitful shoots tend to have large and many clusters. Weather conditions at flowering are very important for fruit set of many winegrape varieties. Cold, wet weather is often associated with reduced berry number per cluster, or small, seedless berries. Fruit set has been shown to be reduced by shaded conditions, and poor fruit set is commonly observed in the centre of dense canopies. Recently, a new disorder has been described to explain this poor set, called **Early Bunch Stem Necrosis (EBSN)**. Individual florets or groups of florets turn brown and fall off the bunch stem during flowering and fruit set. Preliminary studies suggest that this may be due to an accumulation of ammonium, an element toxic to plant tissue at high concentrations.

Berry growth

Final yield is greatly influenced by berry growth which depends on water supply and also the availability of sugars from photosynthesis. Shade in the canopy causes reduced berry growth because photosynthesis is restricted and also the increase in sugar is delayed, leading to a later harvest date.

Disease

Final yield can also be decreased by *Botrytis* bunch rot. This fungal disease occurs principally when it rains near harvest. Again, dense, shaded canopies encourage the development of this disease, as will be subsequently explained.

Inside a dense grapevine canopy, California. (Photo R.S.)

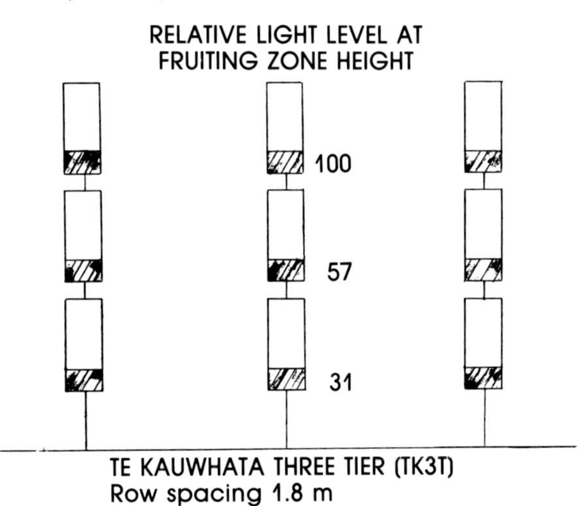

A schematic diagram of the Te Kauwhata Three Tier trellis, designed to investigate light effects on yield and quality.

Shade effects on yield and its components, Cabernet Franc

	Exposed (top)	Intermediate (middle)	Shaded (low)
Yield/vine (kg)	22	8	4
Clusters/vine	169	120	89
Nodes per vine	80	80	79
Percent budburst	109	91	89
Clusters/shoot	1.9	1.7	1.3
Cluster weight (g)	129	68	48
Berry number	92	69	54
Berry weight (g)	1.4	1.0	0.9

These results were obtained with the variety Cabernet Franc at Rukuhia, New Zealand for the 1989 harvest. The Te Kauwhata Three Tier trellis was deliberately designed to study shade effects on yield and quality. Each vine was trained to one of three heights in a vertical plane. Because row spacing is close the low and middle tiers are shaded at the canopy exterior, even though canopies are not dense.

The interior of a dense Cabernet Franc canopy. Note yellow and dead leaves and poorly coloured fruit. Rukuhia, New Zealand. (Photo B.W.)

Among the world's leading vineyards for red wine quality; Chateau Haut Brion, Bordeaux, France. Note the high degree of leaf and fruit exposure. (Photo R.S.)

High vigour and inadequate trellis can cause leaf and fruit shading, with a loss of potential wine quality, California. (Photo R.S.)

Canopy Microclimate and Quality

Introduction

Wines of premium quality are often, but not always, associated with vineyards that have both low vigour and low yield. These observations in both Old World and New World viticulture lead to the concept that low yield and or low vigour are the cause of high quality. We do not believe that low vigour and low yield are the **cause** but are rather **associated** with high quality. Let us explain.

Vineyards with low yields and low vigour typically have open canopies with good leaf and fruit exposure. We believe that improved quality results essentially from the good canopy microclimate, and not necessarily from the low yield and/or low vigour. There are profound implications of this simple explanation. Wine quality can be improved by making canopies more open, especially for vineyards that have high vigour and high yield, and are typically shaded.

Shade and wine quality

Recent studies around the world have demonstrated that shade in canopies causes reduced wine quality. Such results are reported for warm to hot climates; for example in Australia, Chile, California and South Africa and also, in cool climates like New Zealand, Canada, France and Italy. Similarly, the results are common to a range of varieties including Cabernet Sauvignon, Cabernet Franc, Merlot, Shiraz, Chardonnay, Sauvignon Blanc, Traminer and Riesling to name a few. This leads to the development of some general principles for the effect of shade on wine grape composition and, hence, wine quality.

Shade causes:

- decreased sugar levels,
- decreased anthocyanin and phenols in red wines,
- decreased tartaric acid,
- decreased monoterpene flavour compounds,
- decreased varietal or fruit character on the nose and palate,
- increased juice and wine potassium (K) and pH,
- increased malic acid, and ratio of malic to tartaric acids,
- increased 'herbaceous' or 'grassy' characters in the wine, and
- increased incidence of *Botrytis* bunch rot, leading to premature aging of the wine.

In many regions and for many varieties the above list is commonly thought to be associated with overcropping or excessive vigour. However, these problems are more often due to canopy shade. When shade is removed by using appropriate canopy management techniques, these problems are eliminated or reduced and wine quality substantially improved.

This is not to suggest that overcoming shade is the only factor important in producing quality winegrapes. The vineyard should be in balance, with neither too much nor too little leaves and fruit. The vines should experience a mild stress, especially from water. Both these factors are associated with low vigour, and are considered in more detail in the vineyard scorecard.

The Wine Academy
Auckland, New Zealand

Sensory Evaluation of Wine from Canopy Management Trials

For the past two years I have joined a group of industry wine judges at the Te Kauwhata Research Station to taste wines made from grapes grown with various trellising methods. My background in wine tasting includes participation in 20 wine competitions in four countries; tutoring at the Wine Academy, New Zealand's only wine school; and the evaluation of around 3500 wines annually to research my wine writing activities.

For each tasting at Te Kauwhata we were asked to score each wine according to colour density, colour hue, fruit on the nose, fruit on the palate, palate structure, acidity, and overall acceptability. We also used the more familiar 20 point score sheet.

The tasting took place in purpose-built tasting booths and wines were served in numbered XL5 tasting glasses with watch-glass lids. Judges were isolated one from the other. The methods used to conduct these tastings were far more rigorous than any I have encountered in a wine competition.

Generally there was a clear division between the quality of two groups of wines we tasted. The superior group was consistently better in colour, fruit intensity, and structure. These were wines from improved trellis systems like the TK2T, RT2T and Scott Henry. The inferior group of wines we discovered was made from grapes grown on trellising systems which were shaded. The wines were of poorer colour density, often browner, and lacked fruit character. The difference between these wines and the others would have been obvious to any wine drinker. The major surprise, however, was the fact that the better wines were made from vines that yielded more than those which produced poorer quality. It seems that quantity and quality are not mutually exclusive goals, at least for vines of high vigour vineyards.

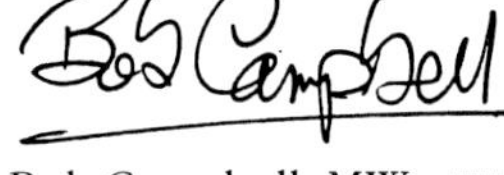

Bob Campbell, MW
Wine Journalist

Bob Campbell, MW

Wine Quality and Yield Results

Hot climate, Variety Shiraz, Angle Vale, Australia

	Dense canopy	Open canopy (Geneva Double Curtain)
Yield (t/ha)	22	27
Wine pH	3.96	3.49
Wine colour density (abs. unit)	2.7	4.3
Wine anthocyanins (mg/L)	280	390
Wine phenolics (abs. unit)	24	37
Wine sensory score (ex 20)	11.9	15.4

Note that the dense canopy increased wine pH, and decreased colour, phenolics and wine sensory score. Despite a higher yield, the more open GDC canopy produced better wine composition and score.

Cool climate, Variety Cabernet Franc, Rukuhia, New Zealand

Treatment	Dense canopy (VSP)*	Open canopy (RT2T)*
Yield (t/ha)	15.8	29.4
Percent bunch rot	19	2
Wine pH	3.40	3.19
Wine colour density	3.9	7.0
Wine anthocyanins (mg/L)	160	165
Wine phenolics	22	24
Wine sensory score (ex 7)	3.5	5.1

Wines produced from the open canopy of the RT2T has lower pH and improved sensory score despite higher yield. (From Smart et al. 1990)

** VSP stands for Vertical Shoot Positioning and RT2T for Ruakura Twin Two Tier training systems. These are described later.*

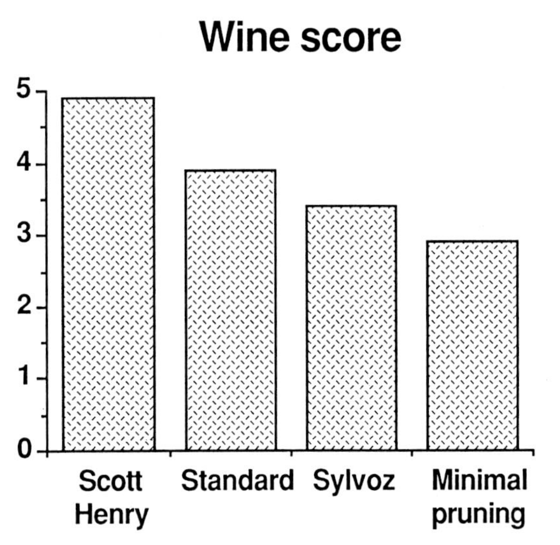

Wine quality scores out of 7 points for Cabernet Franc grown on different trellis systems. Manutuke, New Zealand.

Powdery mildew infection of Sultana berries at the centre of a dense canopy. Australia. (Photo R.E.)

Botrytis bunch rot, New Zealand. (Photo R.S.)

An open canopy with good shoot spacing allows ready spray penetration, RT2T. Rukuhia, New Zealand. (Photo R.S.)

Canopy Microclimate and Grapevine Diseases

Canopy microclimate can affect the incidence and severity of several grapevine diseases. Included are two most important fungal diseases, powdery mildew and *Botrytis* bunch rot. Development of these diseases is encouraged by dense canopies which have high humidity, dry slowly, and are shaded, which factors may be more or less important for the individual diseases. Downy mildew incidence can also be encouraged by dense canopies.

Powdery mildew

Powdery mildew, also called oidium, is native to North America but is now a major disease worldwide. It is caused by the fungus *Uncinula necator*. This fungus may overwinter inside dormant buds or on the vine surface. These sites enable the fungus to infect early shoot growth in spring. Fungal spores are rapidly spread by wind to reinfect other shoots, foliage and fruit. Temperatures of 20-27°C (68-83°F) are optimal for infection and disease development. Rainfall or surface moisture can inhibit development.

Importantly, bright sunlight inhibits germination of spores. Low levels of light, as are found in dense canopies, promote germination. Hence, severe powdery mildew problems are often found within dense canopies.

Botrytis rot

Botrytis bunch rot is caused by the fungus *Botrytis cinerea*. Infection of flower parts can occur before capfall and cause portions of the cluster to fall off. At veraison, once berries start to ripen, the fungus can invade healthy berries through wounds like bird pecks. Compact clusters are particularly at risk from infection because berries under pressure often split and provide an entry point for the disease. Infection can take place of sound berries if they are wet for sufficient periods of time. Compact clusters also encourage berry to berry spread in the absence of injury or wetness. The fungus overwinters on canes, on the bark and in dormant buds. Spore germination and infection requires free water, mild temperatures, high humidity, and susceptible tissue.

Botrytis bunch rot can greatly reduce wine grape yield and quality. Yield is reduced by early loss of clusters and, later, by loss of juice. Crop losses can exceed 30%. Even more important are effects on quality. Wines produced from rotten grapes have off flavours, have unstable colour and are prone to oxidation and age prematurely. Other fungi and bacteria can readily invade *Botrytis* infected clusters which further contributes to off odours.

Under certain environmental conditions, and for certain cultivars, *Botrytis* takes a form known as **noble rot** which concentrates the juice within the berry. Dessert wines such as those of Tokay in Hungary, Sauternes in France and the Auslese type in Germany are produced from fruit with noble rot.

An open canopy makes *Botrytis* control easier than with a dense canopy. Clusters wet by rain or dew dry out more quickly, especially due to increased wind speed in open canopies compared to dense canopies. Since *Botrytis* has developed resistance to some chemicals it is important to reduce dependence on sprays by maintaining an open canopy.

Spray effectiveness

Effective control of many pests and diseases is dependent on chemicals being uniformly and adequately applied throughout the canopy. When canopies are dense, the outer leaf layers intercept most of the applied spray. Penetration is poor to the centre of the canopy, often where the fruit is located. Although such problems can be reduced when spray is forced into the canopy with airblast sprayers, such approaches are not compatible with the move to environmentally sensitive viticulture, as much chemical can be lost to the atmosphere and soil.

Canopies that have low density allow spray material to be more uniformly spread throughout the vine. Also, the design of new spray machines is possible where surplus spray material is collected and recycled.

Dispelling Some Myths

Yield and quality

Perhaps we have created confusion when we have used consecutive statements of increased quality with associated increased yield. It is common belief that increased yield causes decreased quality. This point of view is taken to extreme in parts of Europe where the concept is legislated. Which of these apparently irreconcilable views are correct? Does high yield cause decreased quality, or is there little or no effect?

Let us try to explain with the example of a Cabernet Sauvignon vineyard with moderate vigour in a cool climate. The vines are planted 2 m (6 ft) apart in 3 m (10 ft) rows and trained with only a single foliage wire and no shoot positioning. Some leaf and fruit shading occurs, and the grower believes his vines are too vigorous. If the pruning level were increased from 40 nodes up to 100 nodes the yield may double, but ripening will be delayed. Wine quality would likely be reduced, with less colour and fruit flavour. The traditional explanation would be that high yield **caused** the decrease in quality. Alternatively, the canopy microclimate explanation would be that pruning to more nodes caused more shoots per vine which increased canopy density and this led to increased fruit and leaf shading. This shade contributed to the loss of quality.

If the same vineyard was treated with improved canopy management, say by dividing the canopy, there would be improved fruit and leaf exposure and, hence, improved wine quality. The increased crop could be matured with minimal or no delay, and at the same time the increased node number would have helped improve vine balance.

Note we are **not** saying that high yield always leads to higher quality or that the world's best wines will come from high yielding vineyards. We have however seen many examples in commercial vineyards where yield is being restricted in the belief that quality will be improved. The usual response is that vigour is stimulated, and shade problems are worsened!

Soil effects on wine quality

In the Old World, especially France, much is made of the effect of soil on wine quality. In fact, soil and geological features are often the basis for appellation. Certainly soils do have major effects on vine growth, yield, and quality. This is principally because soil features determine root distribution, thus the supply of nutrients, and, in combination with climate, the supply of water to the vine. So a vineyard on a deep, well structured, loam soil in a high summer rainfall climate will produce vigorous growth, and often high yields. If the canopy management is inadequate, quality will be lower due to excessive shade. Contrast this with growth of a vineyard planted on a shallow soil of sandy texture and containing many stones. Here the vigour and yield will be much reduced. Since shoots will be of low vigour, the canopy will be more open, with good leaf and fruit exposure compared to canopies of the vigorous site, and the wine quality will be higher. From examples like these the idea has developed that soil has a large **direct** effect on wine quality, but we believe these differences can be explained in part by canopy microclimate effects, and are thus **indirect**. Vine growth is largely dependent on water and nutrient supply. Hence application of irrigation and fertilisers, even to an infertile soil, will cause increased vigour. And so some vineyard practices can override soil effects.

Soil conditions of famous, high quality vineyards cannot be exactly duplicated elsewhere. However, the canopy microclimate of these vineyards can be more or less copied. This is the basis of our canopy management procedure.

What we are saying is . . .

Every vineyard in the world cannot be made to reproduce the quality of famous vineyards. This would be to deny the important effects of climate, soil, clone and other factors. However, there are many vineyards which are grown under conditions of excessive vigour, and/or inadequate trellis. Typically, these vineyards produce yields and quality well **below their potential**. And by appropriate canopy management techniques both yield and quality can be simultaneously improved as canopy shade is eliminated or reduced. The following chapters will describe how this can be done.

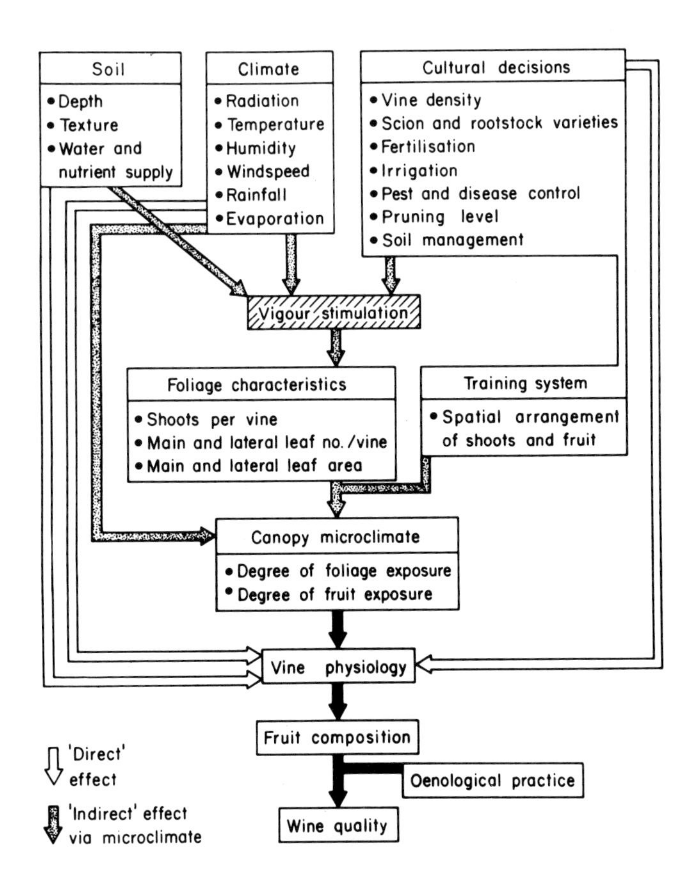

A schematic diagram illustrating factors affecting wine quality. Note the important interaction of vine vigour and training system which determines canopy microclimate, providing for important indirect effects on wine quality. After Smart et al. (1985).

A famous vineyard soil; Chateau Yquem, France. (Photo R.S.)

A high vigour vineyard on a deep fertile soil with a poor canopy microclimate. Oregon, USA. (Photo R.S.)

Quality Assurance in Vineyards

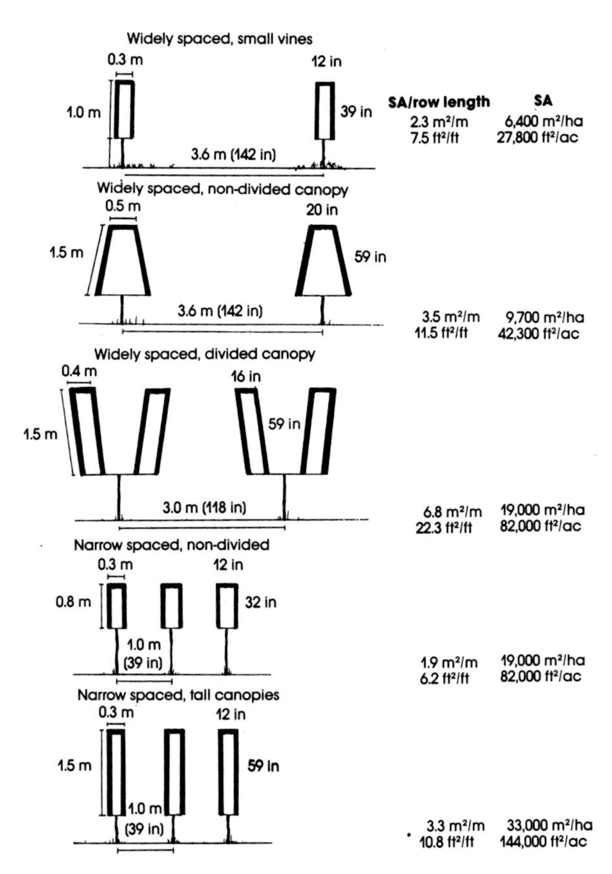

A range of grapevine canopies with different canopy surface area values. Thickened lines indicate exposed surfaces.

Calculating Canopy Surface Area for Widely Spaced, Non-Divided Canopies

Calculations: 1. Calculate row length (L) per unit vineyard area. You need to know the row spacing (R).

	Metric units	*Imperial units*
Formula	L (m/ha) = 10 000 m²/R(m)	L (ft/ac) = 43 560 ft²/R(ft)
Example	R = 3.6 m	R = 11.8 ft
	L = 10 000/3.6	L = 43 560/11.8
	= 2 778 m/ha	= 3 691 ft/ac

2. From your sketch calculate surface area (S) of exposed canopy surface area per unit row length (see diagrams)

Example	S/row length = 3.5 m²/m	S/row length = 11.5 ft²/ft

3. Calculate surface area of exposed canopy per unit vineyard area (SA).

Formula	SA (m²/ha) = S x L	SA (ft²/ac) = S x L
	= 3.5 x 2 778	= 11.5 x 3 691
	= 9 723 m²/ha	= 42 446 ft²/ac

Typical values of exposed canopy surface area

Widely spaced, small vines	6 400 m²/ha	(27 800 f²/ac)
Widely spaced, non divided canopies, large vines	9 700 m²/ha	(42 300 ft²/ac)
Widely spaced, divided canopies	19 000 m²/ha	(82 000 ft²/ac)
Narrow spaced, non-divided canopies	19 000 m²/ha	(82 000 ft²/ac)
Narrow spaced, tall canopies	33 000 m²/ha	(144 000 ft²/ac)

The optimal value is about 21 000 m²/ha (87 000 ft²/ac) for vertically positioned shoots in vineyards which are convenient for manual and mechanical operations. Values much less than this are of vineyards with low yield potential, and higher values are associated with inter-row shading.

The Concept of Quality Assurance

The concept of quality assurance in vineyard management is not widely known or practised. However, for many years quality assurance has been an important feature of winery operation. Why has there been this lack of awareness of quality assurance by vineyard managers?

The likely answer is very simple. Until recently there had not been a system of quality assurance developed for vineyards so that the viticulturist could use these methods to improve vineyard production in terms of both quality and quantity.

Over recent years our group has spent a lot of effort translating research results into practical management tools. We had the following criteria in mind for techniques of vineyard assessment. These techniques should:

- be simple to use,
- be easy to learn,
- be quick to use,
- not require expensive equipment, and,
- give results that are easy to understand and also meaningful.

Our experience when we teach canopy management techniques is that viticulturists can quickly learn vineyard assessment, and are then able to look at their own vineyards with new eyes.

Do I have a problem in my vineyard?

The first step is to establish whether you have a vineyard with a problem canopy. The following pages will introduce you to techniques you can use to answer this question. Most of the techniques are quick and simple to perform. However, for those who require more detailed tests, then these are also available here. Also, as a tip to wine marketers, here is a potential source of wonderful information for back labels! We will introduce you to a range of techniques — determination of canopy area, vineyard scoring, point quadrat, plus a range of other measures of performance. Application of these methods will answer the question — do I have a problem canopy?

Assessment of Exposed Canopy Surface Area

The amount of crop which a vineyard can produce is essentially dependent on its **exposed surface area**. Thus, a vineyard with small vines planted in wide rows can never produce a high yield. This is because there is a low potential to intercept sunlight. Measurement of canopy surface area in association with leaf area can be used to assess canopy density. If there is a large ratio of leaf area to canopy surface area, then most leaves are interior and the canopy is shaded.

Method

Surface area assessment is normally carried out when the canopy is fully grown. This is usually done between veraison and harvest. Simply, the canopy outline is defined as a cross section of the row. If the canopy is 'spiky' you will need to make an arbitrary judgement of the canopy limits. We usually suggest that shoot tips be excluded, and the dimensions include say 90% of the canopy leaf area. This assessment is simplified where canopies are trimmed.

From the sketch calculate exposed canopy surface area per unit row length. Exposed surfaces are those which intercept sunlight and are at the top or on the sides of canopies. Some trellis systems have shaded surfaces — for example the underside of horizontal trellises, or of some steeply inclined systems like the Tatura. These surfaces are rarely exposed to direct sunlight, and should not be counted as exposed. Examples show how the canopy surface area can be calculated. These calculations are for continuous canopies.

Vineyard Scoring

Background

The idea of vineyard scoring was developed by Richard Smart in 1980. During a five month period studying 'famous' and 'not-so-famous' vineyards in the Bordeaux region he found that it was possible to associate the vineyard appearance with its quality reputation. There were a few critical characters to look for to make these assessments.

It was from this idea that the vineyard scorecard for the assessment of potential wine grape quality was developed. These ideas were first published in 1985 after study of different canopy management treatments for Shiraz grapevines at Angle Vale, South Australia. This first approximation scorecard was found to relate well with sensory scores of wine. Subsequently, the scorecard has been through several modifications. The contributions of Diane Kenworthy and Steve Smith are particularly acknowledged. The present scorecard is the fourth version.

The scorecard is used to estimate potential wine grape quality of a vineyard before harvest. This information can be useful for harvest planning. Where it is necessary to mix fruit from different vineyards it is possible to combine vineyards with similar quality potential. Similarly, the scorecard can be useful as a guide to wine grape pricing. Another important application of the scorecard is to monitor changes in canopy management for the vineyard.

Theory

There are two primary groups of vineyard factors that affect wine grape quality. These are to do with **microclimate** and **vine physiology**. Both groups of factors can be visually assessed, hence the concept of a vineyard scorecard. Previous sections in this handbook have emphasised the effects of microclimate on quality. Suffice to add here that improved wine grape quality is associated with vines of moderate vigour, especially where there is no active shoot growth after veraison. Also, the vines should be under slight water and nutrient stress. The vine vigour status just prior to veraison is reflected in leaf size, leaf colour, lateral growth and number of active shoot tips.

The scorecard presented here has eight characters. Each is assessed out of 10 points, for a total 80 points. Three characters relate to the canopy microclimate (canopy gaps, canopy density and fruit exposure) and five to the physiological status (leaf size, leaf colour, shoot length, lateral growth and presence of active shoot tips). Open canopies with evidence of devigouration and slight stress score highest.

It is not presumed that the present scorecard is the ultimate version. Many other characters could be added, e.g. extent of periderm (bark) formation on shoots after veraison, leaf water stress symptoms, etc. Also, some present characters might be eliminated. It may be found that for some varieties in some regions a different emphasis (weight) of characters should apply. Despite these considerations, the scorecard has been found to be a useful method to predict the potential of wine grape quality from vineyards before harvest. The scorecard does differentiate vineyards with different quality reputations, and this has been supported by research studies.

Practice

Scoring can be done by a single scorer, but we have found it useful to have two or more and to average their scores. To assess a canopy takes about two minutes. Have the sun at your back, and score typical canopy sections. It is most important to assess canopies which are representative of the vineyard. Scoring is carried out between veraison and harvest. **Do not use** the scorecard for very low vigour or unhealthy vines, as it will give false values.

Material

All you need are the scorers and a scorecard. Scorers can be easily trained, but some subjective characters, e.g. leaf colour and size, require experience. Plenty of practice is suggested.

Belinda Gould and Richard Smart scoring canopies. Gisborne, New Zealand.

Undisturbed canopy, Rukuhia, New Zealand. (Photo B.W.)

Assessing fruit exposure by brushing leaves to one side. (Photo B.W.)

Mark IV Vineyard Scorecard

This scorecard indicates potential for the production of quality winegrapes

Note: *If the majority of shoots are less than 30 cm (12 in) long, or if vines are clearly diseased, chlorotic, necrotic or excessively stressed, do not score the vineyard.*

A. Standing away from canopy

1. Canopy Gaps (*from side to side of canopy, within area contained by 90% of canopy boundary*).

● about 40%	10
● about 50% or more	8
● about 30%	6
● about 20%	4
● about 10% or less	0

2. Leaf size (*basal-mid leaves on shoot, exterior*).

For this variety are the leaves relatively:

● slightly small	10
● average	8
● slightly large	6
● very large	2
● very small	2

3. Leaf colour (*exterior basal leaves in fruit zone*).

● leaves green, healthy, slightly dull and pale	10
● leaves dark green, shiny, healthy	6
● leaves yellowish green, healthy	6
● leaves with mild nutrient deficiency symptoms	6
● unhealthy leaves, with marked necrosis or chlorosis	2

B. Standing at canopy

4. Canopy density (*leaf layer number from side to side in fruit zone*).

● about 1 or less	10
● about 1.5	8
● about 2	4
● more than 2	2

5. Fruit exposure (*remember that the canopy has two sides normally—that fruit which is not exposed on your side may be exposed to the other side*).

● about 60% or more exposed	10
● about 50%	8
● about 40%	6
● about 30%	4
● about 20% or less	2

6. Shoot length

● about 10–20 nodes	10
● about 8–10 nodes	6
● about 20–25 nodes	6
● less than about 8 nodes	2
● more than about 30 nodes	2

7. Lateral growth (*normally from near where shoots trimmed. If laterals have been trimmed, look at diameter of stubs*).

● limited or zero lateral growth	10
● moderate vigour lateral growth	6
● very vigorous growth	2

8. Growing tips (*of all shoots, the proportion with actively growing tips—make due allowance for trimming*).

● about 5% or less	10
● about 10%	8
● about 20%	6
● about 30%	4
● about 40%	2
● about 50% or more	0

Total point score ______/80 = ______%

Some hints

First, stand away from the canopy with your back against the adjacent row to score the first three characters (canopy gaps, leaf size, and leaf colour).

Canopy gaps: Estimate the proportion of gaps in the canopy. Do not count as gaps the 'holes' which occur between spiky shoots at the edges of the canopy.

Leaf size and leaf colour: These scores require experience with the variety to know what are relatively large or small leaves and healthy, normal coloured leaves. Observe only exterior leaves in a basal to mid shoot position. Be careful not to include lateral leaves in your size assessment as these are generally smaller.

Canopy density: Assess this by putting your face near the canopy fruit zone, and fix your gaze straight ahead. Use your finger to move leaves aside in your line of sight, counting the number of leaves until you reach the other side of the canopy. Count zero for a canopy gap. Record the leaf layer number. Repeat to get a representative number.

Fruit exposure: Develop a mental image of the fruit you can see at the canopy exterior. Then, brush away the leaves, and assess how much fruit is in the canopy and not visible from the outside. Percent fruit exposure is the ratio of the two values. Repeat to get a representative number.

Shoot length: Count the nodes on some representative shoots and average them. Be careful not to unconsciously select the longest shoots. It is a good practice to close your eyes to select shoots so they will be chosen at random.

Lateral growth: Again select some shoots at random and look for lateral growth up and down the shoot. If the shoot has been trimmed, most lateral growth will be near the cut end. Again experience will help you in your assessment. The following guide will also help:

- *Very vigorous lateral growth:* Laterals are growing at most nodes, and the majority of them are longer than five nodes.
- *Moderate vigour:* Laterals are developed at about one third of the nodes on the shoot and most laterals are less than four nodes long.
- *Limited or zero vigour:* Laterals occur infrequently, and normally do not develop more than two nodes long.

Growing tips: Assess all the tips on main shoots and laterals. Tips that are actively growing will always have the blunt apex extending beyond any young leaves on the shoot. Tips which have stopped growing have young leaves which can be folded in front of the growing tip.

Optimum values

The scorecard will give high total points to a canopy which is very open, where vigour is moderate and where the vines are under slight water and nutrient stress.

75-80 points	open canopy with moderate vigour shoots.
About **50** points	dense canopy, but with low to moderate vigour shoots.
20 points	dense canopy, with high vigour shoots which are untrimmed.

Vineyard Scorecard

This summary sheet can be copied for use in your vineyard

	Points	Vineyard and Date			
1. **Canopy gaps**					
• about 40%	10				
• about 50% or more	8				
• about 30%	6				
• about 20%	4				
• about 10% or less	0				
2. **Leaf size**					
• slightly small	10				
• average	8				
• slightly large	6				
• very large	2				
• very small	2				
3. **Leaf colour**					
• leaves green, healthy, slightly dull and pale	10				
• leaves dark green, shiny, healthy	6				
• leaves yellowish green, healthy	6				
• leaves with mild nutrient deficiency symptoms	6				
• unhealthy leaves, with marked necrosis or chlorosis	2				
4. **Canopy density (mean leaf layer number)**					
• about 1 or less	10				
• about 1.5	8				
• about 2	4				
• more than 2	2				
5. **Fruit exposure**					
• about 60% or more exposed	10				
• about 50%	8				
• about 40%	6				
• about 30%	4				
• about 20% or less	2				
6. **Shoot length**					
• about 10–20 nodes	10				
• about 8–10 nodes	6				
• about 20–25 nodes	6				
• less than about 8 nodes	2				
• more than about 30 nodes	2				
7. **Lateral growth**					
• limited or zero lateral growth	10				
• moderate vigour lateral growth	6				
• very vigorous growth	2				
8. **Growing tips**					
• about 5% or less	10				
• about 10%	8				
• about 20%	6				
• about 30%	4				
• about 40%	2				
• about 50% or more	0				
Total score (ex 80)					

Sample Result Using the Scorecard

Cabernet Franc		
Character	**RT2T (open canopy)**	**VSP (dense canopy)**
Canopy gaps	10	0
Leaf size	10	6
Leaf colour	10	6
Canopy density	10	2
Fruit exposure	10	2
Shoot length	9	10
Lateral growth	9	6
Growing tips	10	10
Total ex 80	78	42

The scorecard was used to describe Cabernet Franc vines trained to vertical shoot positioning (VSP) and the Ruakura Twin Two Tier (RT2T). The RT2T gives a score that is superior. These vines had an open canopy and the shoots were devigorated and wine quality was superior. (Smart, et al. 1990)

An actively growing shoot tip, Cabernet Franc. (Photo B.W.)

A shoot tip where growth is slowing or stopped. (Photo B.W.).

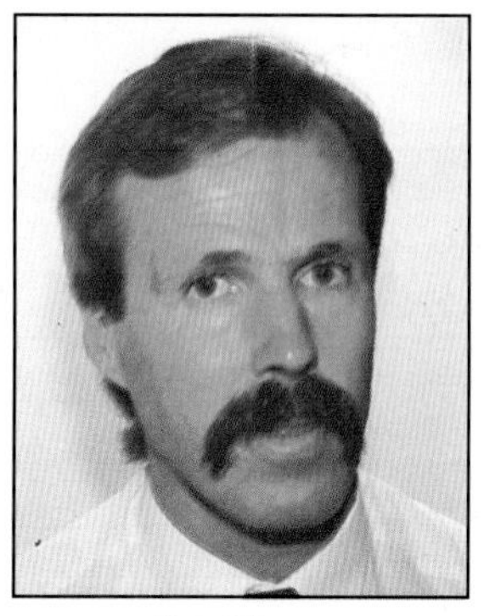

Rob Gibson

Penfolds Wines Pty Ltd
Nuriootpa, South Australia, Australia

Vineyard Assessment of Potential Wine Quality

Penfolds Wines is one of Australia's largest wineries, and uses fruit from a large number of vineyards in different regions to produce a wide range of wine styles and qualities. This range includes the famous 'Grange Hermitage' label, widely regarded as one of Australia's finest dry red table wines. We began vineyard assessment in 1982 with the principal aim of defining the quality potential of each of our vineyards, to facilitate subsequent winery operations. Of particular concern was to identify vineyard sources of Grange Hermitage, so that this supply could be secured against the prospect of growers removing some vineyards. Determining quality potential of the vineyards before vintage allows us to batch grapes in parcels of uniform quality during the intake and crushing processes.

Two field visits are timed at approximately 50% veraison and 10 days prior to the estimated harvest date. We have slightly modified for our own use the original score sheet developed by Richard Smart. Observations are made according to two vineyard scoresheets. These subjective observations are made on at least six vines, and are recorded as averages.

The scorecard used at veraison records leaf condition, presence of growing tips, presence of laterals, average shoot length, periderm development, fruit and canopy shading, and vine size and balance. Two weeks before harvest we assess growing tip presence, leaf condition and fruit exposure, and also assess fruit colour, size, flavour and skin thickness. Chemical analyses are made of sugar, acidity, pH, and K along with berry weight.

Scoring of each parameter varies according to the variety. No general scoresheet has been found effective across all varieties and styles of wines. The score system has been found to give a useful prediction of wine quality; however this aspect is the subject of ongoing development. The quality evaluation is done by a system of batch tracing from vineyard to wine end use and winemakers classification to an extensive range of end product quality grades. Vineyard scoring also helps to explain real differences in quality in terms relevant to vineyard management. These differences could not previously be explained by the simple variety, district and site parameters previously used by winemakers. After seven years experience we found that the relationship of score to wine quality has been improved by separating some varieties and then devising different scorecards for each.

Vineyard assessment has been a useful tool for selection and isolation of premium grape batches for appropriate styles, standardising observation procedures for semi-skilled staff, and focusing on vineyard management practices (e.g. irrigation) to affect relevant parameters (e.g. leaf condition and growing tip presence). Further, it has stimulated the collection of information on vineyard practices and site conditions in premium vineyards. Future research will be conducted in a joint project with the Waite Agricultural Research Institute at the University of Adelaide.

Rob Gibson
Grape Supply Manager

Point Quadrat

Background

A point quadrat is a thin metal rod which is inserted into a canopy. Contacts with leaves and other vine parts are recorded. This technique has been used for many years by research agronomists to study field crop plants. The first recorded use of a point quadrat in vineyards was in 1979–80 by Richard Smart at Angle Vale, South Australia. Since then, the technique has been used in research studies of vineyard canopies elsewhere in Australia, the USA, Canada and other countries. As well, the technique is now used by some commercial vineyards.

Theory

Like a beam of light, the rod passes from the canopy exterior through into the interior. Contact of the rod with parts of the canopy relates to their exposure to sunlight. Using the point quadrat tells us which proportions of the leaves and fruit are exterior or interior in the canopy.

Practice

Normally the rod is inserted into the fruit zone, although any portion of the canopy can be used. Select representative parts of the vineyard to assess. For vertical canopies, the rod is inserted horizontally. For canopies with sloping walls, the rod can be inserted at right angles to the axis of the wall. Where the canopy is very dense or very wide, it is only necessary to insert the rod to the centre of the canopy, although the calculations presented below need to be modified to allow for this (see example). Normally, we suggest 50 to 100 insertions are necessary for representative data. The rod should be inserted at random. Do not look at the canopy before an insertion is made. Often it is desirable to position a guiding board horizontally along the canopy face, and insert the needle at regular intervals (e.g. 10 cm or 4 in) marked along the board. Alternatively, guiding holes may be pre-drilled in PVC pipe at appropriate intervals and the pipe positioned along the canopy wall. After the rod is inserted, record sequential contacts of the vine parts from one side of the canopy to the other. Use **L** for leaf, **C** for cluster, and **G** for canopy gap. You can record shoots **S** if you wish but we normally ignore these.

Materials

A point quadrat can simply be made from a welding rod (bronze or brass) with a sharpened tip. Usually these are long (1 m, 3 ft) and fine (2 mm, 0.1 in dia.). The rod must be rigid. A guide to assist holding the rod steady can be made from a hollow tube 10–20 cm (4–8 in) long of slightly larger diameter than the rod.

Calculations

Sample results are presented from 50 insertions into a low density Cabernet Franc canopy (on RT2T trellis) and 50 for a dense Traminer canopy (with shoots vertically positioned). From the contact information the following descriptions can be readily calculated. Use a highlighter pen to mark off contacts as they are counted.

Percent gaps: The total number of gaps **G** divided by number of insertions (50 here). Multiply answer by 100 to obtain a percentage.

Leaf layer number (LLN): The total number of leaf contacts **L** divided by the number of insertions (50 here).

Percent interior leaves: The number of interior leaves, i.e. not at either surface, divided by the total number of leaves. Multiply by 100 to obtain a percentage.

Percent interior clusters: The number of interior clusters **C**, i.e. not at either surface, divided by the total number of clusters. Multiply by 100 to obtain a percentage.

Richard Smart and Kate Gibbs doing point quadrat analysis of RT2T trellis, Rukuhia, New Zealand. (Photo B.W.)

The point quadrat needle being inserted into a canopy. Note the pre-marked sampling guide. (Photo B.W.)

Richard Smart doing point quadrat assessment of canopies at Schloss Johannisberg, Germany.

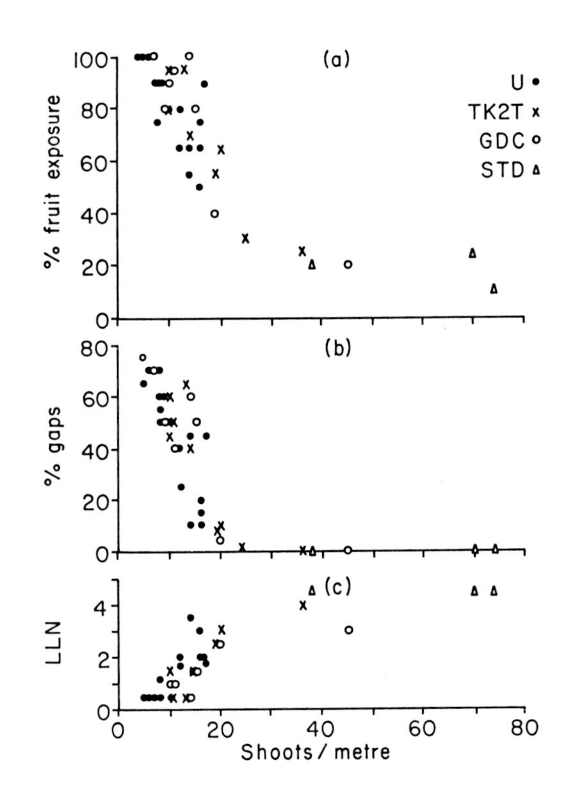

LEFT: The effect of shoot spacing on fruit exposure, canopy gaps and leaf layer number. Less than 20 shoots per metre gives good fruit exposure, sufficient canopy gaps and low values of LLN. After Smart (1988).

Some hints

Make sure you select representative canopies and always work within a fixed zone of the canopy. Do not cheat by looking before you insert the rod. It is easiest to have one person to insert the rod, and another to record. An alternative is to use one person and a portable tape recorder. An insertion can be made and recorded in 10–20 seconds, so about 200 can be done per hour.

Optimum values

Percent gaps should be 20–40%, LLN should be 1.0–1.5 or less, and the percent interior leaves less than 10%, and interior clusters less than 40%.

Sample Point Quadrat Analysis Sheet

A. WITH INSERTIONS FROM SIDE TO SIDE

Low density canopy Cabernet Franc (RT2T)				High density canopy Traminer (VSP)			
1	L	26	LCC	1	L*LCCL	26	LLLLL**
2	LLC	27	LCL	2	LLL	27	LLLLLL
3	LLC	28	G	3	LL	28	LLLLLCLL
4	LC	29	G	3	LL	29	LLLLL
5	L	30	L	5	LLCL	30	LLLL
6	L	31	C	6	LLCLLL	31	LL
7	LCL	32	LL	7	LLCL	32	LLCCLL
8	G	33	G	8	CLLL	33	LCLL
9	G	34	G	9	LCLL	34	LLL
10	LCC	35	LCL	10	LL	35	LCCLL
11	G	36	LLL	11	LL	36	LL
12	L	27	LL	12	LLL	27	LLLLLLC
13	LC	38	L	13	LLL	38	LCLL
14	CL	39	C	14	LCLCL	39	LCCCCLL
15	L	40	G	15	LLLL	40	CLL
16	L	41	C	16	LLL	41	LC
17	CL	42	CLCC	17	LLL	42	CCLLL
18	LL	43	G	18	LL	43	LLL
19	L	44	CC	19	LCL	44	LLCLL
20	L	45	C	20	LLLL	45	LLCLL
21	L	46	LC	21	LCLLLL	46	LLCCL
22	L	47	G	22	LLL	47	LLCCL
23	L	48	G	23	LLL	48	LLLCL
24	G	49	L	24	CLL	49	LLLCLLLL
25	G	50	LL	25	LCC	50	LLCLCL

	Low density	High density
Percent gaps:	13/50 = 26%	0/50 = 0%
Leaf layer number:	43/50 = 0.86	166/50 = 3.32
Percent interior leaves:	4/43 = 9%	73/166 = 44%
Percent interior clusters:	6/23 = 26%	33/40 = 83%

B. WITH INSERTIONS FROM ONE SIDE TO THE CENTRE

For the Traminer data above assume insertions 1 to 25 were made from the west side (so first contact L* is exterior) to the canopy centre, and insertions 26 to 50 from the east side (so last contact L** is exterior). Calculations are:

Percent gaps:	0/50 = 0%
Leaf layer number:	(72/25 + 94/25) = (2.88 + 3.76) = 6.64
Percent interior leaves:	(49 + 71)/166 = 72%
Percent interior clusters:	(12 + 24)/39 = 92%

POINT QUADRAT DATA SHEET

THIS SHEET CAN BE COPIED FOR USE IN YOUR VINEYARD

Vineyard and Date		Vineyard and Date	
1	26	1	26
2	27	2	27
3	28	3	28
4	29	4	29
5	30	5	30
6	31	6	31
7	32	7	32
8	33	8	33
9	34	9	34
10	35	10	35
11	36	11	36
12	37	12	37
13	38	13	38
14	39	14	39
15	40	15	40
16	41	16	41
17	42	17	42
18	43	18	43
19	44	19	44
20	45	20	45
21	46	21	46
22	47	22	47
23	48	23	48
24	49	24	29
25	50	25	50

Percent gaps = ______/50 = ______ %
Leaf layer number = ______/50 = ______
Percent interior leaves = ______/__ = ______ %
Percent interior clusters = ______/__ = ______ %

Percent gaps = ______/50 = ______ %
Leaf layer number = ______/50 = ______
Percent interior leaves = ______/__ = ______ %
Percent interior clusters = ______/__ = ______ %

POINT QUADRAT DATA SHEET

THIS SHEET CAN BE COPIED FOR USE IN YOUR VINEYARD

Vineyard and Date		Vineyard and Date	
1	26	1	26
2	27	2	27
3	28	3	28
4	29	4	29
5	30	5	30
6	31	6	31
7	32	7	32
8	33	8	33
9	34	9	34
10	35	10	35
11	36	11	36
12	37	12	37
13	38	13	38
14	39	14	39
15	40	15	40
16	41	16	41
17	42	17	42
18	43	18	43
19	44	19	44
20	45	20	45
21	46	21	46
22	47	22	47
23	48	23	48
24	49	24	29
25	50	25	50

Percent gaps = ______/50 = ______ %
Leaf layer number = ______/50 = ______
Percent interior leaves = ______/__ = ______ %
Percent interior clusters = ______/__ = ______ %

Percent gaps = ______/50 = ______ %
Leaf layer number = ______/50 = ______
Percent interior leaves = ______/__ = ______ %
Percent interior clusters = ______/__ = ______ %

Effect of canopy density on sunflecks

Length of sunlit rod (cm) for canopies of different shoot spacing.
Traminer, Te Kauwhata, New Zealand. (Rod length 1 m).

Sunfleck number	Open canopy 5 shoots/m	Intermediate 13 shoot/m	Dense canopy 20 shoots/m
1	5.0	0.5	—
2	0.1	0.5	—
3	1.0	2.0	—
4	3.0	0.6	—
5	1.2	0.5	—
6	31.0	—	—
7	0.2	—	—
8	0.8	—	—
9	6.0	—	—
10	0.2	—	—
11	9.0	—	—
12	1.5	—	—
13	8.3	—	—
Total length (cm)	67.3	4.1	0
Percentage sunflecks	67.3	4.1	0
Sunfleck frequency (number per m):	13	5	0

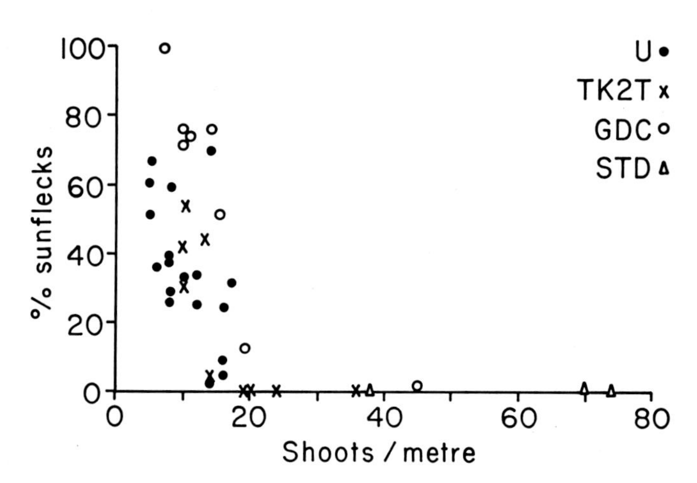

The effect of shoot spacing on sunfleck occurrence. Sunflecks only occur for less than 20 shoots per metre. After Smart (1988).

A good sunfleck pattern underneath a low density canopy on an overhead trellis, Chile. (Photo R.S.)

Sunfleck Assessment

Background
Sunfleck patterns are a good indication of canopy density. They can be assesed in two ways, either in the centre of the canopy or by the shadow pattern on the vineyard floor. The latter procedure is difficult for other than overhead trellises.

Theory
If beams of direct sunlight pass through the canopy they will form sunflecks on a surface beneath the canopy. There is a relationship between the leaf area of the canopy, the angle of the sun and the frequency of sunflecks at the base of the canopy. A dense canopy will intercept more light and have fewer sunflecks than an open canopy. Sunflecks are an indicator of canopy gaps.

Practice
As most grapevine canopies are not overhead, sunflecks are most readily measured at the centre of the canopy, rather than on the vineyard floor. Insert a long (about 1 m or 3 ft) thin rod into the canopy so that it lies in the direction of the row, and in the fruit zone. Make sure the canopy is not disturbed because this will affect the sunfleck pattern. For overhead canopies, the rod can be laid in random positions on the ground. Measure the length of individual sunflecks along the rod. You will need to reposition the rod at about 10 random locations to get a representative measure.

Calculations
Add the total sunfleck length, and count the number of sunflecks. The results can be expressed as:

$$\text{Percentage sunflecks} = \frac{\text{total sunfleck length}}{\text{total length of rod}}$$

Sunfleck frequency = number of sunflecks per unit length of rod (e.g. per metre or foot).

Some hints
To compare canopies, you need to have the sun in about the same position in the sky, relative to the canopy orientation. Measurements are best done when the sun is about at right angles to the canopy face. For vertical canopies, this is when the sun is about 45 to 60° above the horizon. The same row orientation is important for comparisons between canopies or vineyards. Overhead canopies can be analysed with higher sun angles.

Optimum values
Because of the large effect of sun angle and location where measured in the canopy, it is difficult to provide optimum values. For vertical canopies measured in the fruit zone with the sun at about 45°, sunflecks should be about 2–10%.

Light Measurements

It is now possible to buy hand-held meters and sensors to measure light in canopies of grapevines. These can be connected to recording devices to provide averages over time. Also, linear sensors can be used to give averages over distance. We do not recommend their use in commercial vineyards because of their expense and because there are enormous problems with sampling for light measurement, especially in sunny climates. Also, data are difficult to summarize and interpret. If you are not put off by our warning and wish to continue, we suggest that you read the literature on light measurements carefully to avoid pitfalls.

Shoot and Leaf Area Assessment

Background

Both shoot and leaf measurements can help describe some of the critical components that form a canopy. Leaf area measurements are slow to perform, so be sure you really want the information before you commit yourself (or someone else) to do it! However, these measurements are invaluable for fine tuning canopy management. Leaves are the main cause of shade in a canopy, and are important to capture sunlight for photosynthesis. Hence the amount of leaf area and its distribution is of major importance. It determines the degree of shade and also the adequacy of leaf area for fruit ripening.

Theory

Analyses of canopy leaf area are made from leaf areas determined on a per shoot basis. The results can then be expressed as a per vine or per land area basis by counting shoots per vine and vines per hectare (or acre). The following parameters are measured for each shoot:

Main leaf number (N_m) and total area (A_m, cm^2), lateral leaf number (N_l) and area (A_l, cm^2) shoot length (L, cm), main node number (n), cluster number (C) and weight (W, g).

Practice

Shoots should be selected at random including those from both interior and exterior. At least 20 shoots from a vineyard are required for a meaningful average. Shoots are usually sampled just before harvest and they should immediately be placed in large plastic bags and stored in a refrigerator before measurement. Sometimes defoliation has occurred which should be recorded (as % defoliation), i.e. number of nodes without leaves compared to total node number.

Leaf area can be measured with an electronic leaf area meter but these devices are expensive. It can also be estimated by correlation with vein length. The technique we describe using cut discs is both simple and inexpensive. With this technique, the weight of a known number of leaf discs of known area is compared with the weight of leaves, allowing their area to be calculated. To show how this technique is used, we present measurements with Italia, a large-leafed table grape.

Assessing leaf area using the disc technique

A sample of eight main leaves weighed 51.8 g. Twenty discs were cut from these leaves, which weighed 1.2 g. The borer diameter was 1.8 cm.

Area of disc $= \pi r^2 = 3.142\ (0.90)^2 = 2.54\ cm^2$
Area 20 discs $= 50.9\ cm^2$

Weight leaves/weight discs = area leaves/area discs

Then: area leaves (cm^2) = weight leaves (g) × area discs (cm^2)/weight discs (g)
Area of 8 leaves $= 51.8\ g \times 50.9\ cm^2/1.23\ g$
$= 2143.6\ cm^2$
Hence area per leaf (A_m) $= 267.9\ cm^2$

This should be repeated for lateral leaves.

Materials

You will require large plastic bags for the shoots, a top pan balance accurate to 0.1 g and a sharpened cork borer of about 18 mm diameter.

Calculations

Mean main leaf area (cm^2) $= A_m/N_m$
Mean lateral leaf area (cm^2) $= A_l/N_l$
Mean internode length (cm) $= L/n$
Total leaf area/shoot (cm^2) $= A_m + A_l$
Leaf area/crop weight (cm^2/g) $= (A_m + A_l)/W$
Proportion of lateral leaf area (%) $= 100.A_l/(A_m + A_l)$

Optimal values

It is difficult to define optimal values since these can vary greatly with variety and climate. The table on this page is divided into vigour classes, with representative values given for each. In general, values for moderate vigour are preferred.

Optimal Values

	Vigour		
	Low	Moderate	High
Main leaf area (cm^2)	<80	20–160	>180
Lateral leaf area (cm^2)	<25	30–40	>50
Shoot length untrimmed (cm)	50	100	>200
Shoot node number (untrimmed)	<10	15–20	>25
Internode length (mm)	<50	60 – 80	>80
Lateral nodes/shoot	<3	3 – 5	>8
Leaf area/crop weight (cm^2/g)	<5	8 – 12	>20

Sample Leaf Area Measurements

(Averages of 25 shoot samples, 15 nodes long)
Variety Sauvignon Blanc

Measurements		
Main leaf number	N_m	15
Main leaf area (cm^2)	A_m	1936
Lateral leaf number	N_l	16.5
Lateral leaf area(cm^2)	A_l	1153
Shoot length (cm)	L	132.3
Main node number	n	15
Cluster number	C	1.7
Cluster weight (g)	W	119
Calculations		
Total leaf area (cm^2)	$A_m + A_l$	3089
Mean main leaf area (cm^2)	A_m/N_m	129.0
Mean lateral leaf area (cm^2)	A_l/N_l	68.8
Mean internode length (cm)	L/n	8.9
Leaf area/crop weight (cm^2/g)	$(A_m + A_l)/W$	29.5
Proportion of lateral leaf area	$100 \times A_l/(A_m + A_l)$	37%

Leaves removed from a dense VSP canopy to assess their area. Te Kauwhata, New Zealand (Photo R.S.)

Winter prunings removed from a vineyard. Their weight should be 1/5 to 1/10 the weight of crop, or the vineyard is out of balance. (Photo R.S.)

Portion of an RT2T canopy in winter. These canes are about ideal length, diameter, weight, spacing and lateral growth. Rukuhia, New Zealand. (Photo R.S.)

Measurements at Pruning and Harvest

Background

Yield at harvest, and cane and node (bud) counts at pruning are the most informative of all the measurements which can be made to indicate vine balance. Normally, yield is known but few growers take the time during winter to count and weigh canes and record node numbers. This procedure takes only a few minutes per vine and gives important information which can be used for vine management decisions.

Theory

The weight of current season wood removed at pruning (**pruning weight**) gives a good indication of the vegetative growth of the vine during the season. Pruning weight is proportional to leaf area carried on the shoots the previous growing season. It is important to note that pruning weight is reduced by summer trimming. Calculation of **mean cane weight** (pruning weight/cane number) gives a useful indication of shoot vigour.

Measurements of yield and cluster number allow calculation of **mean cluster weight**. If a sample of berries is taken to determine berry weight, then berry number per cluster can also be calculated ignoring the weight of the cluster stem. This will provide an indication of the relative contributions of berry number and berry weight to cluster weight, and, in turn, cluster weight and number to final yield. The **ratio of yield to pruning weight** gives a good indication of the balance between fruit and vegetative growth. Vineyards with high vigour have low values and overcropped vines have high values.

Practice

All measurements are done on a per vine basis. Measurements at harvest and pruning can be made on 10 or more vines. These vines should be representative of the vineyard or of blocks within the vineyard. At pruning, the first operation is to count canes per vine. For normally pruned vines, very small canes (say five nodes or less) can be ignored. After pruning, measure pruning weight and nodes retained. Similarly at harvest clusters are counted from sample vines and berry weight is determined from a 100–200 berry sample.

Materials

Pruning weights are easily measured with a hand-held spring balance with 5–10 kg (10–20 lb) capacity, accurate to 0.1 kg (0.2 lb). The prunings are bundled with a string and hook, and attached to the balance. Fruit yields at harvest can be measured with a top pan mechanical balance or a tared spring balance with a cradle for the fruit bin. The calculation of berry number per cluster assumes the cluster stem weight is zero. In fact it is often about 10%, so this calculation is useful only for comparative purposes.

Calculations

Mean cane weight = pruning weight/cane number
Mean cluster weight = yield/cluster number
Berries per bunch = cluster weight/berry weight
Yield/pruning ratio = yield/pruning weight

Optimal values

Optimal values will vary with variety and climate. Representative values are given for low, medium and high vigour vines. Indices which apply to balanced vines are typically those listed for vines with moderate vigour.

	Vigour		
	Low	Moderate	High
Mean cane weight (g)	<10	20-40	>60
Yield/pruning weight (trimmed canes)	>12	5-10	<3

Sample Measurements for Cabernet Franc on Different Training Systems Rukuhia, New Zealand (1986–87)

Measurements (per vine)	VSP (vigorous)	RT2T (moderate vigour)
Pruning level (nodes)	48	152
Cane number	54	167
Pruning weight (kg)	3.7	3.5
Yield (kg)	7.6	29.7
Bunch number	77	299
Mean berry weight (g)	1.63	1.70
Calculations		
Mean cane weight (g)	69	21
Mean cluster weight (g)	99	99
Berries per bunch	61	58
Yield/pruning weight	2.1	8.5

Gristina Vineyards Inc.
Long Island, NY, USA

Pruning Measurements to Improve Wine Quality

Our goal is to maintain a full trellis with leaves and fruit well exposed to light. The vineyard is cordon (spur) pruned, vertically trained, and shoots are supported with two pairs of catch wires. Our vines are spaced 9 × 8 ft (2.7 × 2.4 m). We have no irrigation and receive 20–24 in (500–600 mm) rainfall during the growing season. Fruit yields are about 3 t/ac (8 t/ha). Annually, we assess our vine balance by inspecting the same group of sample vines. They represent about one percent of each vineyard block. In early winter, we prune the sample vines and then count the canes and weigh them. Mean cane weights are calculated for each vine. We compare measurements of previous years for our sample population of vines and factor in the present year's rainfall and any nitrogen additions.

Our mature vines have prunings that weigh from 1.1 to 1.6 kg. Mean cane weights commonly range from 27 to 73 g/cane. We feel that 10–13 shoots per metre of row (25–30 shoots per vine) gives us a full canopy without excess shade on leaves and fruit. Our ideal vine is composed of 30 canes, that weigh 45 g, to give a total pruning weight of 1.4 kg. We use this analysis to determine if we are leaving the proper amount of buds during pruning, i.e. growing the ideal number of shoots per vine.

In 1988/89 our Chardonnay sample population wood weight profile looked like this: <0.9 kg (17%), 0.9–1.1 kg (33%), and >1.1 kg (50%). This told us that we could achieve full yields of about 9–10 t/ha on 50% of our vines, less than full crop of about 6 t/ha on 33% of our vines, and small crops of 0–4 t/ha on 17% of our vines. For the 50% of our vines that are smaller than needed for full crops, we use a combination of growing fewer shoots, fruit-thinning, nitrogen application, mulching and replanting to increase vine size for full crop production.

Peter Gristina
Vineyard Manager

Larry Fuller-Perrine
Winemaker

Peter Gristina and Larry Fuller-Perrine.

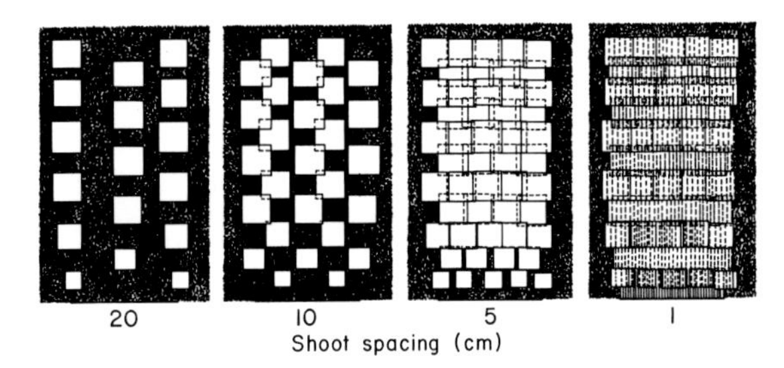

Schematic diagram to show the relationship between canopy shading and shoot spacing.

This spur spacing will give good shoot spacing. Cabernet Franc, Rukuhia, New Zealand. (Photo R.S.)

A canopy showing good leaf and fruit exposure and a good proportion of gaps, due to correct shoot spacing. Cabernet Franc on RT2T, Rukuhia, New Zealand. (Photo R.S.)

Improving Canopy Microclimate

Definition of an Ideal Canopy

This is a useful stage to consider what an ideal canopy looks like, because the next sections in the handbook discuss methods to achieve this.

To be practical, there is never likely to be one ideal canopy, but rather a range of them. For example, modifications may be made for specific varieties in certain climates. However, the general guidelines presented here should apply to both white and red varieties and for a range of climates, and are a useful first approximation.

The values presented have been taken from a recent review paper by Smart et al. (1990) and the reader is referred there for details and justification of values (see Further Reading). This series of descriptors and indices are presented as a **wine grape canopy ideotype**—that is, a description of an ideal canopy. These characters can serve as a checklist to further assess your vineyards. To assist this, we have indicated the relative importance of each character in altering wine quality.

Winegrape Canopy Ideotype

Character	Optimal value	Importance for wine quality	Justification
Canopy characters			
Row orientation	north-south	low	Promotes sunlight interception especially for sunny regions.
Ratio of canopy height: alley width	~1:1	high	High values lead to shading of canopy bases. Low values mean inefficient sunlight interception.
Foliage wall inclination	Vertical or nearly so	high	Underside of inclined canopies is shaded.
Fruit zone location	Near canopy top	mod	Promotes yield although fruit phenol levels can be excessive.
Canopy surface area (SA)	About 21 000 m^2/ha (92 000 ft^2/ac)	high	Low values associated with low yield potential. Higher values not possible without excess shade.
Ratio of leaf area: canopy surface area (LA/SA)	<1.5	high	Indicates canopy density. Low values desired.
Shoot spacing	~15 shoots/m canopy (4.6 shoots/ft)	high	Gives about ideal canopy for moderate vigour vines in vertical canopies. Values for non-positioned canopies can be higher.
Shoot and fruit characters			
Shoot length	10-15 nodes 0.6-0.9 m(2-3 ft)	high	Short shoots cannot ripen fruit property — long shoots contribute to shade and high pH.
Lateral development	Restricted - say less than 5-8 lateral nodes per shoot	high	Excessive lateral growth indicates high vigour and causes shade.
Ratio leaf area: fruit weight	~12cm^2/g (6ft^2/lb) range 6-15 cm^2/g (3-8 ft^2/lb)	high	Low values are inadequate to ripen fruit, high values cause high pH.
Ratio of yield: canopy surface area	1-1.5 kg/m^2 (0.2-0.3 lb/ft^2) for cool climates, 3 kg/m^2 (o.6 lb/ft^2) is a likely maximum figure for hot, sunny climates	high	Exposed canopy surface area required to ripen grapes.
Ratio yield: pruning weight	5-10	high	Indicates vine balance — low values mean excessive vigour, high values mean overcropping.
Growing tip presence	nil	high	Encourages fruit to ripen.
Cane weight	20-40g (0.7-1.4 oz)	high	Indicates desirable vigour level and vine balance.
Internode length	60-80mm (2.4-3.1 in)	mod	Values vary with variety. Indicates desirable vigour level.
Pruning weight	0.3-0.6kg/m canopy (0.2-0.4lb/ft)	high	Lower values indicate canopy too sparse, high values indicate excess shade, for vertical canopies. Higher values possible for non-divided canopies.
Microclimate characters			
Proportion of canopy gaps	20-40%	high	Higher values cause sunlight loss. Lower values lead to excess shade.
Leaf layer number (LLN)	1.0-1.5	high	Higher values associated with excess shade, lower values with sunlight loss.
Proportion exterior fruit	50-100%	high	Shaded, interior fruit are inferior for winemaking.
Proportion exterior leaves	80-100%	high	Shaded leaves cause yield loss and fruit quality defects.

Cabernet Franc canopy on Ruakura Twin Two Tier trellis. Note a blue board is positioned behind the canopy, emphasizing the leaf and fruit exposure, and canopy gaps. Rukuhia, New Zealand. (Photo R.S.)

Cabernet Franc canopy on vertical shoot-positioned trellis. Note limited fruit exposure and yellow leaves due to shading. (Photo R.S.)

WINEGRAPE CANOPY IDEOTYPE

THIS SHEET CAN BE COPIED FOR USE IN YOUR VINEYARD

Character	Optimal value	Your vineyards and date				
Row orientation	north-south					
Ratio of canopy height: alley width	~1:1					
Foliage wall inclination	About vertical					
Bunch zone location	Near canopy top					
Canopy surface area (SA)	About 21 000 m^2/h (92 000 ft^2/ac)					
Ratio of leaf area: canopy surface area (LA/SA)	< 1.5					
Shoot spacing	~ 15 shoots/m (4.6 shoots/ft)					
Shoot length	10-15 nodes, 0.6-0.9 m length (2-3 ft)					
Lateral development	5-8 lateral nodes per shoot					
Ratio of leaf area: fruit weight	~12 cm^2/g					
Ratio of yield: canopy surface	cool climate 1-1.5 kg/m^2 (0.2-0.3 lb/ft^2) hot sunny climate to 3 kg/m^2 (0.6 lb/ft^2)					
Ratio of yield: pruning weight	5-10					
Growing tip presence after veraison	nil					
Cane weight	20-40 g (0.7-1.4 oz)					
Internode length	60-80 mm (2.4-3.1 in)					
Pruning weight per length canopy	0.3-0.6 kg/m (0.2-0.4 lb/ft)					
Proportion of canopy gaps	20-40%					
Leaf layer number (LLN)	1.0-1.5					
Proportion exterior fruit	50-100%					
Proportion exterior leaves	80-100%					

Low vigour bush vines due to water stress. South Africa. (Photo R.S.)

Severe leaf roll virus infection, Hunter Valley, Australia. (Photo R.S.)

High vigour vineyard on deep, fertile soils. Sonoma Valley, California. (Photo R.S.)

Devigoration or Invigoration?

Background

Few vineyards we see are in perfect balance with their environment and trellis system. An important part of canopy management is to determine the vineyard balance and whether **invigoration** or **devigoration** (or neither) is required. Vineyards of low vigour and high vigour are regarded as out of balance. Appropriate management strategies to alter balance can be incorporated in the overall canopy management program.

Measurements of vineyard balance

Previous sections have identified methods which assess vine balance. With practice, many of these assessments can be done quickly and reasonably accurately by visual inspection alone. Those attributes which can be assessed as a guide to vine balance include: leaf size, shoot length, lateral growth, the presence of growing tips after veraison, shoot and internode length, main and lateral leaf area, and the proportion of lateral leaves. Other measurements can include yield and pruning weight, plus their ratio, and, by calculation, pruning weight per unit length row, and mean cane weight. Further, the results of point quadrat and visual scoring analyses are useful.

Common causes of low vigour imbalance

Throughout the world, water stress is probably the most common cause of low vigour. Where water supplies are available there are well-developed irrigation technologies to prevent this stress. Alternatively, improvements to the size and health of the root system by appropriate soil management can often help supply water, especially where root depth is limited by physical or chemical conditions of the soil. Mineral deficiency is not a common cause of low vigour. Moreover, this problem can be identified by soil or tissue analyses and corrected by fertiliser application. Pests and diseases can commonly cause devigoration but control measures are often available. Of particular interest are some virus diseases that devigorate vineyards. The effect of these diseases cannot be overcome, and often the vineyard should be replanted.

In general, low vigour vineyards do not have too dense canopies, but yield is low. Low vigour vineyards are uneconomic in many places in the world. If the cause of the low vigour cannot be overcome, there is often not a need to contemplate canopy management changes. However, if vigour can be improved, for example by irrigation, then changed canopy management may be essential. For example, a simple trellis system may be adequate for low vigour without irrigation, but totally inadequate for increased vigour with irrigation, where shade is the typical consequence.

Common causes of high vigour imbalance

Newly established vineyards often have problems associated with excessive vigour. Many reasons for this excessive vigour can be traced to application of new vineyard technology, especially since the Second World War. These include improved pest, disease and weed control through use of agricultural chemicals, improved planting material free of major virus diseases, improved diagnosis for, and application of, fertilisers, and improved soil preparation and management.

Combined with these practices has often been the selection of vineyard sites where the soils are deep and fertile. Also, in some instances, summer rainfall is high or irrigation is generous. Shoot vigour is often stimulated by training shoots vertically upwards.

Many vineyards we inspect have vines that are out of balance due to excessive vigour. The existing trellis systems are often inadequate to accommodate this vigour. As a result, excessive shade in the canopy occurs, and yield and quality are below their potential. Canopy management techniques should be used to devigorate and improve vine balance.

Improving vine balance

There is no need to further discuss methods of vineyard invigoration since this technology is generally well known. Rather, the emphasis in the following sections will be on methods of devigoration. This is the common problem, and the technology is not as well understood.

High Vigour, the Vicious Cycle

The vicious cycle

High vigour vineyards on restrictive trellis systems can, very quickly, get into a **vegetative growth cycle** which favours shoot growth over fruit production. The excess shade associated with these vines depresses the individual processes that affect yield, so crop weight per shoot is low. In response to the low yield, vegetative growth is further encouraged, and the shade problems increase, so the vicious cycle of vegetative growth is perpetuated. Often high yields occur in the first year with vigorous varieties on high potential sites, and with simple trellis systems. After the first year though, yields decline as shade worsens. Not only is yield lost but quality is also reduced due to the shade.

Breaking the cycle

The vegetative growth cycle can be broken if shade in the canopy is reduced. Once this happens, the cycle can be converted into one which favours balanced fruit and shoot production. Increased light in the canopy stimulates the processes which increase yield per shoot. Then, competition with the cluster restricts shoot growth, and canopy density can be reduced.

There are two methods used in vineyards to break the vegetative growth cycle. Often the most effective is a change in trellis to reduce shade. The following sections will explain how. Devigoration, the second technique, is detailed here.

Devigoration by root zone management

For vineyards in areas with low summer rainfall, irrigation management is a powerful method to restrict shoot growth. Water stress can be gradually increased after fruit set when irrigation is withheld. Active shoot growth can then be halted by veraison. For vineyards in summer rainfall areas, it is more difficult to induce water stress. In these vineyards, stress is applied through increased root competition. Two options are available—grass down the inter-row alley, or increase the density of vines.

Grass cover crops: These compete with the vines for water and nutrients and limit vine root development, especially near the surface of the soil. As well, cover crops improve vineyard access and limit water runoff and erosion. However, there are disadvantages. In some situations the water and nutrient competition can cause too severe vineyard stress. Further, the chance of a spring frost is greater with a cover crop compared to a compact, bare alleyway. Different grass species will use different water amounts. This allows the grower to choose different degrees of competitiveness. Generally, it is desirable to avoid clover-grass mixes because clover may make available too much nitrogen for the vines.

High density planting: This has been given a lot of attention in recent New World plantings. Again the idea is to increase competition below ground for water and nutrients, to restrict shoot growth. Research and commercial experience especially in Europe indicate that high density vineyards only achieve limited devigoration, particularly for low potential soils. For high potential soils, the opposite effect can occur. The closely spaced vines must be pruned to restricted node numbers and vine vigour and shading typically are increased.

Root pruning: This technique has been studied in South Africa for the control of vine vigour. In general, root pruning stimulates new root growth, and there is likely to be a difference between rootstocks in their response. Repeated and severe root pruning has been shown to reduce yield and vigour of irrigated vineyards in South Africa. Roots should be pruned after harvest. Presently, there is insufficient experience to encourage this technique for commercial vineyards.

Grassing down this vineyard has helped reduce vigour in a summer rainfall area. Waiheke Island, New Zealand. (Photo R.S.)

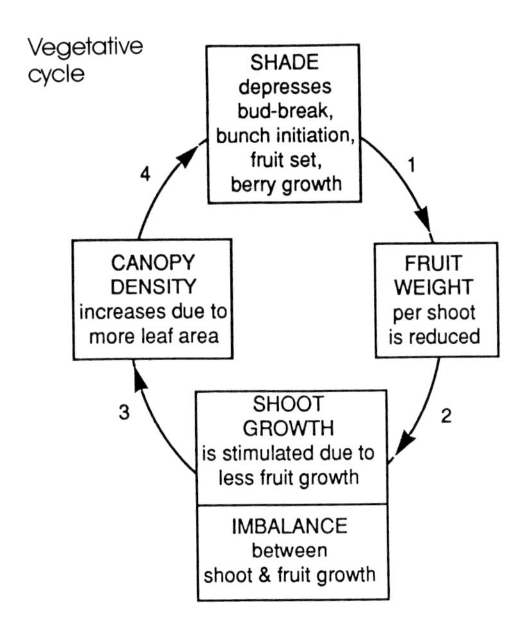

'Vegetative cycle'. Canopies tend to become more and more shaded and vegetative, leading to yield and quality reductions. Typical of vineyards on high potential soils with an inadequate trellis.

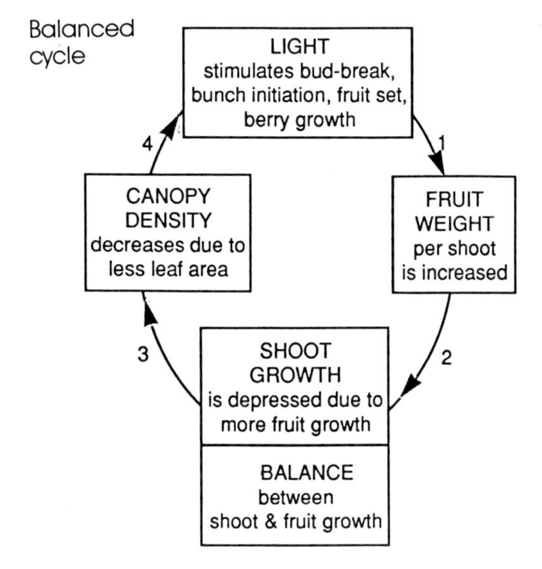

'Balanced cycle'. Light stimulates shoot fruitfulness leading to balance between shoot and fruit growth. Typical of vineyards trellised according to vigour.

A dense canopy and vigorous growth with close in-row vine spacing on a high potential site. Stellenbosch, South Africa. (Photo R.S.)

Vineyard rows planted far too closely together causing excessive shading. England. (Photo R.S.)

Eben Archer

University of Stellenbosch
Stellenbosch, South Africa

Effect of Vine Spacing on Growth, Yield and Quality

Under the influence of European viticulture, many viticulturists in the New World like South Africa, New Zealand, Australia, California and Canada believe that grape and wine quality can be improved by using closer vine spacings. Yet examples exist where such spacings achieve the opposite effect. On the other hand, examples can also be found where positive effects on yield and quality are obtained, underlining the European principle of smaller vines with smaller crop per vine. The tantalizing problem is to choose the correct vine spacing for a specific locality. We have had research in progress since 1979 to better answer this question.

It soon became apparent that **optimum vine spacing is dictated by the potential of the soil**. Soil potential describes the ability of the soil to induce a certain growth performance in the vine and the higher the soil potential, the higher is the capacity of the vine to grow. Generally, low potential soils need closer vine spacings to optimize yield and quality, while soils with higher potential need wider vine spacings.

The second principle of vine spacing found in our experiments is that **between-vine spacing can exert a more important effect on yield and quality than between-row spacing**. A too narrow spacing between vines in the row has marked effects on shoot crowding and within-canopy shade, especially when medium and/or high potential soils are used. Rows should not be spaced wider or narrower than the height of the trellising system used so that the ratio of trellis height:alley width equals about 1:1. The third principle is that **vines should be far enough apart to provide enough cordon space for a balanced bud load**. On a well balanced vine the budburst of collar buds and growth of water shoots is minimized, resulting in lower requirements for desuckering, topping and leaf removal. Such vines are naturally devigorated, resulting in improved canopy densities.

Close vine spacing on low potential soils induces a higher level of interplant competition: the root systems are small and the corresponding above ground growth is restricted. This competition between closely spaced vines creates optimal canopy densities which, in turn, induces a more favourable canopy microclimate. This is especially true during the early part of the season. Later in the season, a higher plant water stress is obtained with more closely spaced vines and because of the negative effect this has on stomatal conductance, a higher leaf temperature is observed. The higher plant water stress is obtained when no or little irrigation is applied, and is induced in closely spaced vines through a higher root density, which causes a higher rate of soil water depletion.

Vine spacing has a marked effect on both yield and quality, mainly through the important effect on canopy density. When low potential soils are used for viticulture, more closely spaced vines augment yield and wine quality, and within-row spacing can be as low as 1.0 m (3 ft). When high potential soils are used, within-row spacing should not be less than 2.5 m (8.3 ft). High vigour inducing practices such as irrigation and nitrogen nutrition on high potential soils often dictate between-vine spacing of more than 3.0 m (10 ft). The closer the spacing between vines, the more necessary it is to use better adapted trellising systems to obtain optimal canopy density.

Archer.
Eben Archer
Senior Lecturer in Viticulture

Devigoration by Shoot Management

Pruning level controls vigour—the 'big vine' effect

Growers often complain about excessive vineyard vigour but few realize that this is often a direct consequence of their own actions in pruning too severely in winter. When vines are severely pruned, only a few shoots are produced in spring and these shoots grow fast and long—that is, they are vigorous. This principle has been well understood for many years, see for example the excellent discussion of this by Winkler and co-authors in *General Viticulture*. A wider appreciation of this principle began with the use of mechanical pruning in Australia and other countries in the 1970s. More recently, the practice of minimal pruning, as developed by the CSIRO in Australia, has shown that shoot vigour is very effectively controlled by light pruning. Most shoots stop growing by flowering and shoot length may be as short as 15 cm (6 in) on minimally pruned vines. In our view, the devigoration on minimally pruned vines can be excessive. It is more desirable to have shoots still growing after fruit set, although at a slow rate. Combined with a single trim which cuts the shoot to the desired length (10–15 nodes), a sufficiently devigorated shoot will have minimal lateral regrowth and limited requirement for future trimming.

It is easy to understand why shoot growth is restricted for lightly pruned vines. Early shoot growth depends on stored reserves within the permanent parts of the vine. There are fewer reserves per shoot when these are spread among many shoots compared to when there are a few shoots. This limitation of stored reserves restricts early shoot growth for lightly pruned vines. As well, there is little lateral shoot development. Leaf area per shoot is thus restricted. Once fruit set has occurred, the developing cluster competes with the shoot tip for products of photosynthesis and since leaf area per shoot is already limited, the growth rate of the shoot remains slow.

The **big vine** theory simply says that high vine vigour can be controlled by pruning to more nodes, and spreading these nodes out so that there is low shoot density and the canopy is not shaded. The big vine approach requires that as more nodes are left, there must be an increase in the space allotted to the vine to avoid shade. Typically this will only apply to vines that are cordon trained. Remember, for an open canopy, there should be about 15 shoots or nodes/m (5 shoots/ft) of canopy. This will produce fruitful shoots, which in turn reduce shoot vigour.

How many nodes are required to achieve devigoration?

A rule of thumb is that 30–40 nodes are required per kg pruning weight (14–18 nodes/lb), so a vigorous vine with 3 kg (6.6 lb) pruning weight requires 90 to 120 nodes. At 15 buds per m (5 shoots/ft), the vine canopy should be 6 to 8 m long (20–26 ft)! If the vine is trained to a divided trellis, with 2 m of canopy per m row, then vine spacing in the row should be 3 to 4 m (10 to 13 ft). These spacings are wider than those normally found in vineyards, so when a vineyard is converted to big vines, normally every other vine, or perhaps two out of every three vines, need to be removed. The vines that remain are trained to occupy the vacant space. When a high potential site is anticipated, new vineyards can be planted with wide within-row spacing. Special care needs to be given to vine training in the first two years to develop cordons quickly. This will ensure that production commences as soon as possible. For example, Chardonnay vines planted 2 m apart on RT2T on a high potential site at Rukuhia, Hamilton, were trained to fill 8 m (26 ft) of cordon at the end of the second growing season! This was achieved because of the high natural fertility of the site, applied nitrogen, and careful training during the first two years.

Shoot orientation also controls vigour

Shoots growing upwards are more vigorous that those trained downwards. The devigorating effect of downward shoot growth is used to advantage for some training systems, to be described (Geneva Double Curtain, Scott Henry and Ruakura Twin Two Tier B form).

A famous 'big vine'. The Winkler vine, University of California, Davis. (Photo R.S.)

Wild vines covering forest trees. These vines are never pruned (by man)! Virginia. (Photo R.S.)

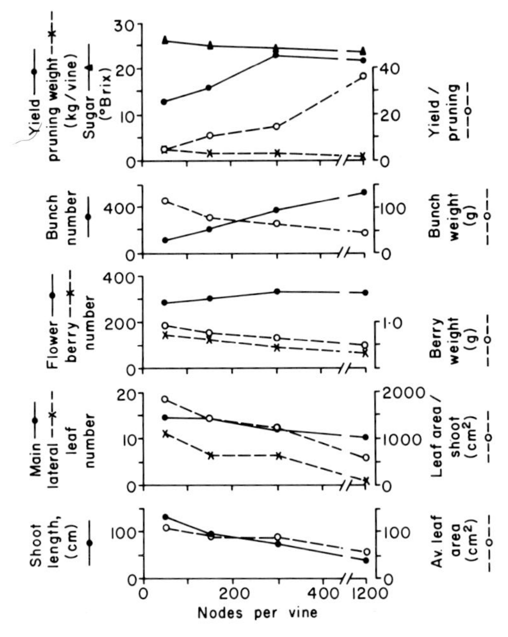

Results of a pruning trial with irrigated Shiraz vines. As more nodes are retained, the vigour of individual shoots is decreased, being shorter and with fewer and smaller leaves. Also, the balance between yield and pruning weight improves. After Smart and Smith (1988).

Shoot Growth Measurements
Traminer at Te Kauwhata (sampled 19 December 1989, 26 days after flowering, and averaged over four trellis systems)

	Big vines, 96 nodes	Small vines, 48 nodes
Shoot length (mm)	1021	1144
Node number	24.6	24.1
Shoot diameter (mm)	13.6	13.9
Leaf area, node 4 (cm^2)	85.3	95.0
Growing point number/shoot	5.7	7.7
Mean internode length (mm)	41.4	47.0

Extensive sucker or water shoot growth due to vines being too severely pruned. Water shoots are typically non fruitful, and so contribute to shade but not to yield. (Photo R.S.)

Shoot tips from 'small' and 'big' (on right) Cabernet Franc vines taken 29 November 1989. Note shoot growth slowing for the 'big' vines. Rukuhia, New Zealand. (Photo B.W.)

Waikato University
Hamilton, New Zealand

Effect of Node Number on Shoot Devigoration

In a current study, I am measuring the response of shoot growth to different node numbers per vine but similar nodes per metre cordon with Traminer at Te Kauwhata, the so-called 'big vine' effect. Vines were originally planted at 1.5 m (5 ft) by 3 m (10 ft) and during winter 1987 every second vine was removed in some plots to allow a comparison of 'big' vines at 3 m (10 ft) spacing with 'small' vines at the original 1.5 m (5 ft) spacing. Cordons of big vines were extended to fill the gaps. The big and small vines were pruned to 96 or 48 nodes per vine, respectively, for non-divided canopies vertically shoot positioned (VSP) and the divided canopies of U, TK2T and GDC. Shoot growth measurements during the 1989–90 season show that big vines had reduced shoot vigour.

Big vines had fewer water shoots—of the total shoots there were 36% water shoots on big vines compared to 49% on small vines. Shoots of big vines grew less vigorously than shoots of small vines. This was seen in the shorter internode lengths of big vine shoots compared to shoots on small vines. Measurement of shoot diameter showed that big vine shoots were also thinner than small vine shoots. The rate at which leaves were produced was unaffected, with similar numbers of main leaves per shoot between the two vine sizes. However, the mature leaf area on big vines was smaller than for those of the small vines. Big vines had fewer growing points than small vines, due to less lateral growth on big vine shoots.

The big vine effect was obvious for each trellis system studied. There were also some differences between trellis systems. The U (or lyre) trellis and the VSP produced the most vigorous shoots compared to other trellises, perhaps due to less severe trimming in previous seasons. These measurements support the big vine theory. Increased node number per vine, but with little change in nodes per metre of row, does lead to shoot devigoration.

Chelle Gandell.
Michelle Gandell
M.Sc. student

Michelle Gandell. (Photo D.J.)

Canopy Management—How To Do It

The importance of canopy management has become clear from the preceding sections. These have shown the effect of canopy microclimate on yield and quality, and how to diagnose canopies with problems. Maybe you now realise there is a need for canopy management in your vineyard. Or you may have a new vineyard planned and want to get the canopy management correct from the start. Now you have to decide which canopy management technique to use.

Canopy management techniques can be divided into three categories — **pre-plant**, **temporary** and **permanent** solutions for established vineyards. Pre-plant decisions include site selection and plant density. Temporary solutions must be repeated each year. These include shoot thinning, trimming and leaf removal. Permanent solutions include devigoration and trellis changes.

Pre-plant canopy management—site assessment

It must seem strange to think about canopy management before a site is selected, let alone planted. However this is often the best time to do it. Presently there are many vineyards which have canopy problems that are difficult to control because there was insufficient consideration made of canopy management at the initial stages.

It is necessary to evaluate the **potential** of any site, i.e. determine what will be the vigour of the vineyard. The major factor that affects this potential is water supply. This evaluation requires detailed soil inspection to determine what the potential depth of the root zone will be. It is a good idea to dig pits with a back hoe and have an experienced person look for features which will inhibit root growth, i.e. compacted layers. Soil tests should be done to check for problems, e.g. nutrient deficiencies or unsuitable pH. The soil can be categorized as:

- **high potential**, with potential root zone more than 1 m (3.3 ft) deep,
- **medium potential**, with 0.5–1 m (1.6–3.3 ft), and
- **low potential**, with less than 0.5 m (1.6 ft) depth.

There is a large interaction between climate, which affects water supply, and soil potential. These factors combined dictate site potential. Low summer rainfall and high evaporation will promote water stress. This is especially evident on low potential soils. Such sites will need to be irrigated to avoid excessive water stress. Alternatively, vineyards planted on high potential soils in climates with adequate summer rainfall will be very vigorous, and irrigation will not be necessary. Warm, sunny climates also favour vigorous vine growth more so than cool, cloudy climates. Other clues about site potential can be obtained from adjacent vineyards or agricultural plantings on similar soils. Are the plants vigorous, with high yields? What have been the irrigation responses? Another clue is to observe pastures on the soil, when do they stop growth? If the grass stops growth early in the summer, then you may need to irrigate your vineyard, but if the grass is always green, then it is unlikely you will need to irrigate. Also, there is a good chance your vineyard will have excessive vigour.

Now you have determined the soil potential, the effect of climate, and whether or not you will irrigate. From these you should have an idea of your site potential. This information will influence your planting decisions! The following is a guide.

High potential sites: You can anticipate high vineyard vigour if the site has a high potential. Then allowance should be made for a more elaborate training system which will accommodate the vigour and vine spacing to match, e.g. wide rows (3–3.6 m, 10–12 ft), and wide spacing within the rows (2 m or 6 ft at least).

Low potential sites: On low potential sites, row and vine spacing can be smaller. There will not be a need for a complex trellis. Thus, row spacings down to 1.5 m (5 ft) and vine spacings to 1 m (3 ft) may be appropriate.

Before these decisions are made, read further about the trellis which will best suit your site and plant at the appropriate spacings.

Shiraz vineyard showing excessive water stress, having been planted on a shallow soil in a low summer rainfall region. Barossa Valley, South Australia. (Photo R.S.)

Studying vineyard root distribution allows for a better understanding of factors determining soil potential. Stellenbosch, South Africa. (Photo R.S.)

Richard Thomas and Richard Smart determining vineyard site potential in a soil pit. Sonoma Valley, California. (Photo K.G.)

Desuckering in a South African vineyard to reduce canopy density. Ideally this should be done earlier and only non-fruitful shoots removed. (Photo R.S.)

Problems caused by trimming a non-positioned canopy. Many shoots pointing into the row are cut too short. Australia. (Photo R.S.)

Problems caused by trimming a non-positioned canopy. Any trimming will remove exterior, healthy leaves and expose interior, shaded and non-functional leaves. California. (Photo R.S.)

Temporary Solution Approaches to Canopy Management

Sometimes we call temporary solutions **'band-aid viticulture'**, emphasizing that such practices need to be carried out each season as part of vineyard management. They will not cure a vineyard of canopy problems in the sense that the problem disappears forever. Rather, this type of approach needs to be used each growing season, and the effect lasts only for that growing season. Band-aid approaches to canopy management include shoot thinning, trimming and leaf removal from the bunch zone. Sometimes these approaches may be the best ones to use in the vineyard, as they are often cheap, and can be cost effective. But remember, you will have to repeat the operation for each and every year of the vineyard's life.

Shoot thinning

Vines commonly push more nodes than the 'count' nodes left at winter pruning. These shoots develop from buds at the base of spurs, or out of old wood, and they are commonly called **water shoots**. Water shoots can be a high proportion of total shoots produced, over 50% when a high vigour vine is severely pruned. Water shoots are often not fruitful, and thus cause shading but do not contribute to yield. Shoot thinning or desuckering is carried out in spring to remove water shoots, especially from the cordon with spur pruning. This is quickest to do when shoots are only about 15 cm (6 in) long. Leaving the operation until later makes it more difficult to see whether or not the shoots are fruitful, and where they come from. For an open canopy, thin to about 15 shoots/m (5 shoots per ft) of canopy. However, removal of some shoots means that remaining ones will grow more vigorously encouraging lateral shoot development, so shade may not be reduced. Further, fruit set can be reduced. We have seen evidence of this for Chardonnay and Traminer. Shoot thinning may also lead to reduced production in the following year, due to primary bud necrosis. So try out shoot thinning on a limited area first.

Shoot trimming

Trimming is sometimes called **slashing** (Australia, California) or **summer pruning** (Europe) and consists of cutting off shoot tips during the early summer. This is normally done by a tractor-mounted machine and a wide range of configurations are available. Trimming typically will begin soon after flowering and may need to be repeated during the rest of the season. When vigorous shoots are trimmed, lateral regrowth occurs from near the cut position which makes frequent trimming necessary. Trimming is widely practised in Europe, New Zealand and South Africa, but is not common in traditional Australian or Californian vineyards.

Shoots are normally trimmed to 10–20 nodes length, the figure depending on training system used and variety. Those varieties with large bunches require longer shoots with more leaf area. Normally about 10 nodes is the minimum shoot length to provide sufficient leaf area to ripen fruit adequately. Trimming is almost impossible to do where shoot positioning is not practised since shoots grow in all directions. Thus, shoots protruding into the row are often cut too short, and those lying along the top of the canopy may not be trimmed at all. Early trimming done at flowering when just the shoot tips are removed (topping) can increase fruit set in some situations. Studies in Italy by Solari et al. have shown that trimming 25 days after flowering did not affect yield, fruit sugar and tartaric acid, though malic acid was slightly increased. Juice pH and K were lowered.

Trimming late in the season should generally be avoided, since lateral regrowth may be stimulated during fruit ripening. Also avoid trimming which removes healthy green leaves at the canopy exterior, and exposes yellow leaves from the canopy interior. These yellow leaves are ineffective in photosynthesis.

MAF Technology Ruakura Agricultural Centre
Hamilton, New Zealand

Early Bunch Stem Necrosis and Shoot Thinning

Early bunch stem necrosis (**EBSN**) is a disorder that can severely reduce fruit set in grapes. Symptoms of EBSN are dead sections of the cluster near flowering which often fall off. In extreme situations the whole cluster can be affected. Although recently described by David Jackson in New Zealand and Bryan Coombe in Australia, this disorder is not new. It now seems that EBSN has occurred for many years but was overlooked, or loosely described as 'poor set', or attributed to botrytis infection. The specific cause of the disorder is unknown. Recent research has concentrated on the role of nutrients and, in particular, my work has indicated that toxic levels of ammonium are involved. Some varieties are more susceptible than others and there are clonal differences.

A study with Traminer, a variety that appears particularly susceptible to EBSN, showed that early shoot vigour affected the incidence of EBSN. Shoot vigour was altered by thinning the shoots to either 50 or 10 shoots per vine (vine spacing 3 × 1.5 m) five weeks before bloom. Heavily thinned vines had more vigorous shoots than the lightly thinned vines. This was seen in a two-fold increase in the number of lateral shoots for vines heavily thinned compared to lightly thinned vines. These vigorous shoots had twice the incidence of extreme EBSN. Of the clusters present at bloom, only about a third remained four weeks later, whereas two thirds remained on the shoots of lightly thinned vines. The cluster loss occurred because they were totally affected by EBSN. These results have important implications for canopy management. If shoots are thinned before bloom to reduce canopy density the yield can be greatly reduced because of increased severity and incidence of EBSN.

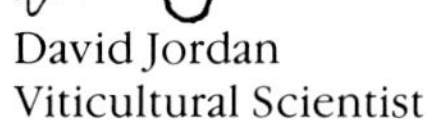

David Jordan
Viticultural Scientist

David Jordan (Photo G.H.)

Problems of poor fruit set due to EBSN are common in high vigour, shaded canopies like this one. California. (Photo R.S.)

EBSN on Chardonnay vines induced by high shoot vigour. South Australia. (Photo R.S.)

Leaf removal can be mechanized to reduce cost.

Cabernet Franc canopy of VSP before leaf removal, Rukuhia, New Zealand. (Photo B.W.)

Same canopy following hand leaf removal. The zone to the right of the trunk has been excessively leaf stripped, that to the left is correct with about 60% fruit exposure. (Photo B.W.)

Leaf removal in a Graves vineyard, France. (Photo R.S.)

Leaf Removal

Leaf removal from the cluster zone is a traditional practice in many European vineyards. Typically it is done at about veraison to reduce the incidence of *Botrytis* bunch rot.

Theory

Leaf removal from the fruit zone opens up the canopy which increases the exposure of the clusters to sunlight and wind. The clusters then dry more quickly after dew and rain. Also exposing fruit to the sunlight can improve the composition for winemaking. Normally, only one or two leaves are removed per shoot, so that there should remain sufficient leaf area to ripen the fruit. Note leaf removal from the cluster zone does not reduce canopy shade elsewhere. Consequently this practice is considered remedial and a permanent change in management of the canopy should be adopted to reduce the canopy density.

Practice

Leaf removal can be carried out any time between fruit set and veraison. The ideal time is 2-4 weeks before veraison. With very early leaf removal, lateral leaves can grow in the fruit zone which may mean the operation has to be repeated later in the season. Avoid sudden exposure of shaded fruit in mid-summer as this can lead to fruit sunburn. Early in the season, the fruit is less prone to sunburn.

Lateral shoot removal in the bunch zone can also be effective in improving fruit exposure. This is easiest to do early in the growing season, say before flowering. The operation needs to be done by hand, and is most effective for medium to high vigour vineyards. A small reduction in fruitfulness may result but canopy density in the fruit zone is markedly reduced.

It is not necessary to remove all leaves from the fruit zone or to expose every part of every bunch. Retain sufficient leaves so that about 60% of the fruit is visible. It is difficult to do leaf removal in a canopy that is not shoot positioned because the fruit is often at the centre. Leaf removal can be done manually requiring up to 50 hrs/ha (20 hrs/ac), or by machine which takes about 3 hrs/ha (1 hr/ac).

Benefits

Leaf removal can be a very cost effective operation, especially if done by machine. Fruit rot can be considerably decreased. This is due to both spray penetration and cluster microclimate being improved. Thus, considerable loss of yield can be avoided. For example, studies with Sauvignon Blanc in New Zealand have shown that leaf removal increased yield by 40% over control vines. Californian studies have also shown a small increase in yield due to improved shoot fruitfulness after several years of leaf removal.

Leaf removal also improves fruit composition. Sugars can be slightly increased and pH and titratable acidity reduced, especially because of reduced malic acid content. Reduction in titratable acidity is the most common response. Also, the colour and phenol content of black grapes can be improved. Other benefits from leaf removal are reduced herbaceous or grassy wine characters, and increased ripe fruit characters.

Robert Mondavi Winery
Oakville, California, USA

Mondavi Experiences with Leaf Removal

We began the commercial application of leaf removal from vineyards in 1982. The original efforts focused on the reduction of *Botrytis* bunch rot. In our Napa Valley climate this form of canopy modification is very effective, as warm dry weather often follows summer storms. Additionally, we use leaf removal as a tool to change the characters and the type of wines which we produce for both red and white varieties. We find that leaf removal is very effective when the canopy is excessively dense and fruit shading gives undesirable chemical composition or aromatic traits to the vines. The positive effects of leaf removal on wine style surfaced immediately upon our first experiences with Sauvignon Blanc, which had canopies opened to avoid rot. The general goal of the Fume Blanc wines at the Robert Mondavi Winery is to reduce the herbaceous (grassy, weedy) component of the wines. We prefer to enhance the fruit characters of melon or mature fruit and the floral complex characters.

We have worked with the University of California, Davis over three years in commercial and small scale vineyard trials. These trials sought to establish the most effective time and severity of leaf removal to modify the above wine characters. We now recommend for our climate and vine canopies to do leaf removal in the early period of as close to berry set as is practical. This timing reduces fruit sunburn and increases the desired effects on chemical composition and aromatic and flavour characteristics. We now believe the results of sunlight penetration into the canopy and the fruit zone to be a dosage response in that the longer the exposure the greater the effects.

It is less clear which severity of leaf removal is optimal. Currently we use integrating light bar meters to monitor exposure at the cluster zone through the growing season in an attempt to establish the proper exposure. We find the Richard Smart rule of thumb of 1 to 1.5 leaf layers between the fruit and the sun is a good working rule for canopy density. This is especially true for a variety such as Sauvignon Blanc and in certain cases of Cabernet Sauvignon grown on fertile soils. Total canopy exposure also has important effects on wine character. We never remove leaves and trim shoot tips in a combination which will result in fewer than 15 functional leaves per shoot. We do not see an effect on yield from leaf removal, although a slight increase in bud fruitfulness resulted in some trials. However, yield was reduced in trials where leaves were removed before complete berry set.

Phillip Freese
Director of Winegrowing

Virginia Polytechnic Institute and State University
Winchester, Virginia, USA

Results of Leaf Pulling Research in Virginia

The macroclimate of northern Virginia typifies much of the mid Atlantic region. Rainfall during the summer and harvest averages about 4 in (10 cm) per month and the mean temperature of July is around 20° C. It is not uncommon to harvest white cultivars at 20° Brix, 3.5 pH, 6 g/l titratable acidity, and with high potassium. Fruit rot development frequently determines harvest date. Rot organisms include *Botrytis cinerea* and other fungi, acetic acid-forming bacteria, and wild yeasts.

Several canopy management research projects were begun in 1986 to explore options to improve fruit quality in this environment. One project evaluated the impact of selective leaf pulling on fruit rot incidence and fruit composition. This experiment was conducted over three seasons in two vineyards and with two cultivars, Chardonnay and Riesling. Vines in one vineyard (high trained) were trained to a high bilateral cordon at 6 ft (1.8 m) and shoots were positioned downwards. Vines in the other vineyard (low trained) were trained to a lower bilateral cordon at 4 ft (1.2 m), and shoots positioned upwards. Treatments were control (no leaves pulled) and leaf pulling (2 to 3 leaves per shoot on average) from around fruit clusters, applied about three weeks after full bloom.

Point quadrat analyses and water-sensitive spray paper obliteration (from airblast sprayer simulating pesticide application) both illustrated increased fruit zone exposure by leaf pulling. Leaf pulling also increased sunlight penetration into the fruit zone, particularly in the high trained vineyard, as determined by light measurements. In fact, some fruit were sunburnt in the high trained vineyard with both cultivars. The principal benefit of leaf pulling was a reduction in fruit rots, up to 20% reduction in incidence and 10% reduction in severity. Reductions in fruit rot were greater and more consistent with Riesling than with Chardonnay. The visual reduction of rot due to leaf pulling has usually been accompanied by reduction in metabolites (glycerol, acetic acid, and ethanol) of decay organisms determined by lab assays of juice. Effects of leaf pulling on other aspects of fruit chemistry have been less consistent. Soluble solids and pH have been generally unaffected. Titratable acidity (TA) was typically reduced by up to 1.0 g/l at harvest in the high trained vineyard. Reductions in TA were less pronounced in the low trained vineyard. Increases in potentially volatile monoterpenes have accompanied leaf pulling with Riesling in both vineyards. A further benefit of leaf pulling that we have noted, but not quantified, is the increased speed that fruit can be harvested by hand where leaf pulling is done.

Tony K Wolf,
Viticulturist

Bruce W Zoecklein,
Enologist

Phil Freese

Tony Wolf

Bruce Zoecklein

Improved Trellis Systems

Modifying Trellis Systems

Often the best solution to improve a dense canopy is to modify the trellis system. This should provide a **permanent** solution. However, other practices such as vine removal, trimming and leaf removal may also need to be incorporated into the canopy management. The need for these depends on vineyard vigour and training system. Often an improved training system can be **retrofitted** to existing vineyards. Only occasionally does the vineyard have to be replanted.

Characteristics of improved training systems

There are several common features of improved trellis systems. They all provide, to a greater or lesser extent:

- increased canopy surface area, because the canopy is divided,
- decreased canopy density, because shoots are devigorated from more nodes per vine, and improved shooting spacing,
- increase possibilities for mechanization of trimming, leaf removal, harvesting and winter pruning,
- improve yield and quality, due to less shade in the canopy,
- better spray penetration due to less dense canopies, and
- lowered incidence of *Botrytis* and powdery mildew.

Classification of trellis systems

There are many trellis systems in use around the world. We do not attempt to discuss them all here. Rather, we will highlight the most popular systems now being evaluated in commercial vineyards in the New World. These seem to meet most requirements to improve yield and/or quality and to facilitate mechanisation. However, other systems not discussed may also meet some of the same goals.

The systems to be described in detail are:

VSP — Vertical shoot positioning (commonly called the 'standard' in New Zealand)
U — or lyre trellis
TK2T — Te Kauwhata Two Tier
GDC — Geneva Double Curtain
SH — Scott Henry
RT2T — Ruakura Twin Two Tier
Sylvoz
MP — minimal pruning (called **MPCT** in Australia)

Choice of trellis systems

Each system will be subsequently described in detail in the following pages. Here we give a few ideas which are a guide to some important factors governing choice of trellis systems. Most important considerations are vineyard potential, variety vigour and row spacing. The suitability of trellis systems to vineyards or varieties of different vigour, and to different row spacings are shown in the tables. In general, vertically divided canopies only can fit in narrow rows. More complex training systems like the RT2T are best suited to high vigour sites and wide rows. There are other considerations as well. These include costs of installation, labour requirements for shoot positioning and other vine management practices current availability of machinery like harvesters, the suitability of present vine form, posts and wire position for conversion, etc.

Vineyard retrofitting — some hints

It is normally possible to convert a vineyard to a new training system during winter pruning. This may or may not require the addition of new posts, cross pieces, strengthened end assemblies or wires. Make sure that you do not trim too hard in the previous growing season. Decide on your new training system early so that you keep sufficient shoot length to form new cordons. Often in conversion to a new trellis system, long canes are laid down to form new cordons. To ensure good bud break, only use well matured, dark coloured canes of medium vigour. Choose the canes which were in the sun the previous year. Disbud any areas where bud break is not required. A spray of 3% hydrogen cyanamide solution applied three weeks before budburst helps give an even budburst. (Check if this chemical is registered for use on grapevines in your locality). For vineyards which are out of balance and heavily shaded, it is normal to recover the cost of installation of the new trellis with the first year because of improved crop, without allowances for the improved quality. This increased yield should be maintained into the future. You may need to replace or strengthen end assemblies if post height or wire number is increased.

Do it right

Many growers are tempted to take short cuts in conversion to a new trellis system and its management. We urge you not to do this. Do not take short cuts on the specified number of wires, or stated dimensions. Yield and quality improvements will not result when the systems are not properly installed. If the systems **do not look right, they will not work right.**

Classification of Trellis Systems

Shoot Orientation	Canopy division: Not divided	Divided horizontally	Divided vertically	Divided both horizontally and vertically
All up	VSP	U	TK2T	RT2TA
All down	Sylvoz	GDC	—	RT2TC
Both up and down	MP	—	SH	RT2TB

Trellis System Suitability to Vine Vigour

Vigour: Low	Medium	High
VSP	VSP, SH	GDC
SH	U, TK2T	Sylvoz
Non-positioned	GDC	RT2T
	Sylvoz	MP

Trellis System Requirements for Row Spacing*

Row spacing: Narrow <2 m (6 ft)	Intermediate ~3 m (10 ft)	Wide >3.6 m (12 ft)
SH	U, GDC	RT2T
TK2T	MP	Non-positioned
VSP	Sylvoz	

* This is the narrowest spacing at which the system can normally be used. Often machinery width is critical.

Vertical Shoot Positioned Trellis (VSP)

Background

The vertical shoot positioned (**VSP**) trellis is widely used in France, Germany and New Zealand to name only a few countries. It is an example of a non-divided canopy. It was developed or adopted in places where there is a high risk of fungal diseases, as it was important to keep the foliage off the ground, to make the vines easy to spray and trim.

Description

Normally the fruiting zone of the vine is about 0.9-1.2 m (3-4 ft) above the ground, and shoots are trained vertically upwards. The shoots are held in place by at least four foliage wires, in pairs. Shoots are trimmed at the top and sometimes along the sides. Shoots are typically 15-20 nodes long. Vines can be either cane or spur pruned. In New Zealand, where cane pruning is popular, the vines are commonly pruned to four canes (about 60 nodes), and there is a vertical gap of about 15 cm (6 in) between two parallel fruiting wires. Two canes are trained in each direction, one on each wire. When vines are spur pruned there is normally a single cordon.

Benefits

VSP vineyards always look attractive when they are neatly trimmed. They are easy to machine harvest, and with some modifications can be converted to machine winter pruning. Because the shoots are uniformly trained, all the fruit is in one zone and the shoot tips are in another. This makes mechanical operations easy, like leaf removal, bunch zone spraying, and trimming. Shoots positioned vertically upwards suits the growth habit of the majority of *vinifera* varieties.

Disadvantages

The VSP system is prone to shade so is unsuited to high vigour varieties and high potential sites. Shoot density is normally high. For example, with vines spaced 2 m (6 ft) apart in the rows and pruned to 60 nodes, there will be about 30 shoots per metre (9 shoots per ft). Remember, the recommendation is 15 shoots per metre. VSP requires taller posts than occur in some vineyards. If used on wide rows of 3 m (10 ft) or more, canopy surface area is below optimum. For example, with canopies 1.2 m (4 ft) high and 0.4 m (1.3 ft) wide on 3 m (10 ft) rows, canopy surface area is only 9 300 m^2/ha (41 000 ft^2/ac). Consequently the VSP system is not particularly high yielding, and shade-induced quality losses are common. In fact in high vigour sites, VSP training may reduce yield and quality compared to non-positioned canopies, as the latter has a greater surface area.

Management guidelines

Normally three passes are needed through the vineyard for shoot positioning. The total time required is about 18 hrs/ha (8 hrs/ac). Shoot positioning is quickest with moveable foliage wires, as these can be moved out to catch shoots then lifted up. The first tucking takes place about flowering, and the final before veraison. Trimming by machine takes about 1.5 hrs/ha (0.6 hrs/ac). Shoots are normally trimmed to about 1 m (3 ft) long with the top 0.3 m (1 ft) extending past the top pair of foliage wires. Shoots may not stand erect if this distance is greater than 300-400 mm (12-16 in).

Material requirements

A post 1.8 m (6 ft) out of the ground is recommended, although 1.5 m (5 ft) is a minimum length. Each row requires 1 or 2 fruiting wires, and four foliage wires. Posts are normally 7.5-10 cm (3-4 in) diameter and set 6-8 m (20-26 ft) apart in the row.

Retrofitting

The system is easy to fit to existing vineyards. It should not be used for rows that are less than 1.5 m (5 ft) apart. Sometimes post extenders can be fitted cheaply to support foliage wires. The VSP system can be an improvement over non-positioned canopies for low to medium potential sites, but should not be used for high vigour sites.

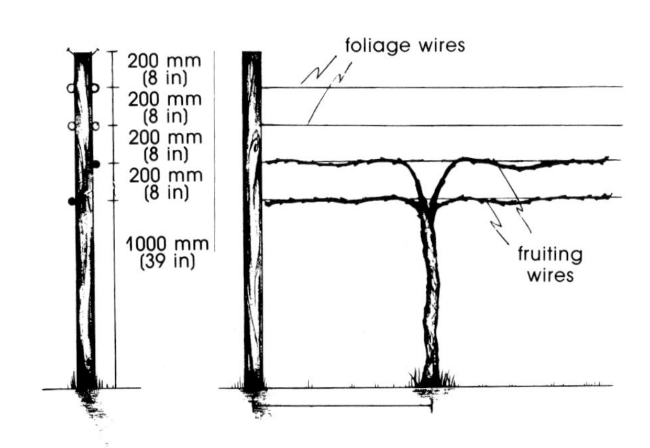

End view and side view showing pruning to four canes for vertical shoot-positioned trellis.

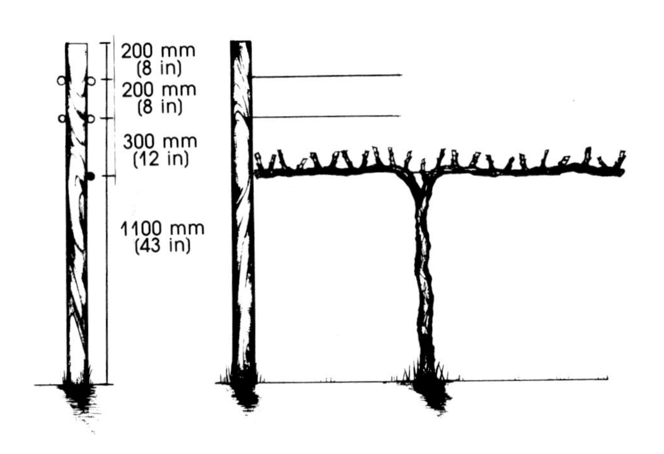

End and side view for VSP trellis showing spur pruning.

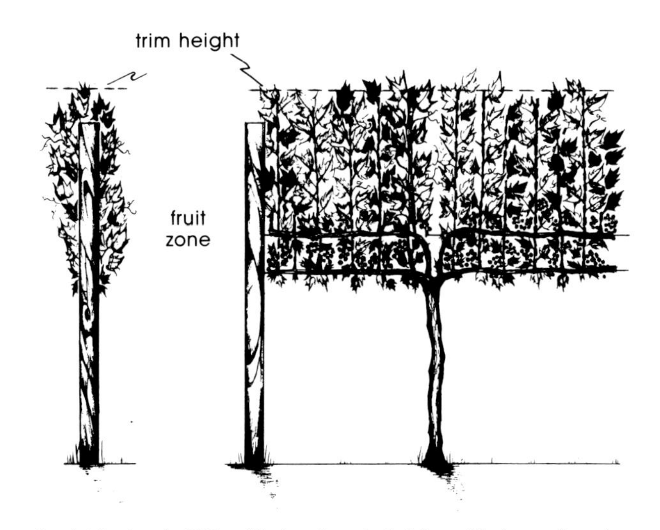

End and side view for VSP trellis showing trim height and fruit zone location, cane pruning.

A well trimmed vertically shoot-positioned trellis at Gisborne, New Zealand. (Photo R.S.)

A poorly trimmed and positioned VSP canopy, South Australia. Only one pair of movable foliage wires was used, and trimming was carried out too late on these vigorous vines. (Photo R.S.)

Zack Berkowitz

Domaine Chandon
Yountville, California, USA

Vertical shoot positioning
Can be fun,
Quality improvements
And an extra ton.

Adoption of Vertical Shoot Positioned Canopies in Carneros

What is a good canopy management system for vines with moderate growth characteristics? Vertical shoot positioning (VSP) has been a very successful canopy management system for our Pinot Noir and Chardonnay sparkling wine vineyards in the cool Carneros region of California.

Typically, our vines have a low to moderate amount of growth. Crop to pruning weight ratios are between 5:1 and 8:1. Shoots grow to about one metre in length and leaf area to fruit ratio is between 7 and 10 cm²/g. In short, these vines do not have the common vigour problem.

Three years of trellis and spacing trials have led to the conversion of 120 hectares (300 acres) of standard non positioned California canopy to VSP. In addition, all plantings since 1989 include this system. The typical dimensions of our vines are: fruit wire at 90 cm (35 in) and catch wire positions at 120 and 150 cm (48 and 60 in). One pair of catch wires is moved at about 50 cm (20 in) of growth. These same wires are moved to the second position at about 100 cm (40 in) of growth. After this last shift of the catch wires shoots are trimmed by machine. Trimming does not remove much foliage but assures that the wall of foliage remains constrained by the wires. Little or no regrowth occurs after trimming. The resulting shoots have about 15 nodes. Total labour to move wires twice and remove them before pruning is about 20 to 25 hrs/ha (8 to 10 hrs/ac).

VSP has resulted in many improvements over non-positioned canopies. Water status, as measured by pressure chamber and infra-red thermometer, is improved with VSP and by late in the season the shoot positioned blocks have more photosynthetically active leaves. All machine-related operations are easier with VSP. Machine harvesting has been much cleaner with no breakages. Fruit and foliage exposure is better, resulting in improved pesticide applications, i.e. more material is hitting the target. Even shoot trimming is more uniform with all shoots oriented in the same direction. We now have renewed interest in mechanical pruning. A leaf removal implement was tried on positioned vines with much success although we feel that leaf removal is not required on these vines. From a practical standpoint, positioning shoots is simple and makes other operations easier and more effective.

Yields have improved in every trial comparing positioned vines to those without positioning. VSP vines have more clusters per shoot and have larger berries. Differences of up to 1.5 t/ha (0.6 t/ac) are typical. Fruit analyses show higher titratable acidity and lower malate at the same sugar content. VSP wines are preferred for their improved aroma intensity and mouthfeel. Wines from positioned vines have softer acids and a delicate fruit character — both important for sparkling wines. This is truly a case of improved quality and quantity.

Visitors to our VSP blocks often ask why we do not grow our rows closer together than 3 m (10 ft). Trials with row spacing of 2 m (6 ft) have been quite successful. Main and lateral leaf growth is reduced and yield and quality continue to improve as rows are planted closer. Close row spacing and VSP are the future for the vineyards of Domaine Chandon.

Zack Berkowitz,
Vice President, Vineyard Operations

Scott Henry Trellis (SH)

Background

This trellis system was developed in the early 1970s by Scott Henry, an Oregon grapegrower and winemaker. Scott and his wife Sylvia have vineyards on fertile soils, which resulted in very vigorous vine growth. This situation led them to develop a system to improve yield and quality. The Scott Henry is a vertically divided training system. It is a variation of the VSP system, but half the shoots are trained upwards and the other half downwards.

Richard Smart saw this system in Oregon in 1983, and was impressed with the concept of vertical canopy division, similar to the Te Kauwhata Two Tier principle. The idea was brought back to New Zealand and evaluated. The system was seen to have considerable potential in New Zealand since the conventional cane pruned VSP system could be readily converted. The Scott Henry system has since been modified to facilitate downward shoot positioning, and also to allow spur pruning. The system is now being widely evaluated in Australia and California as well as New Zealand and Oregon. For a while the system was termed the 'Smart Henry' in Oregon, but now is widely known as 'Scott Henry'.

The Scott Henry system is set up like a VSP cane-pruned vine. Rather than shoots from both cane positions being trained vertically up as in VSP, for the Scott Henry the shoots from the top cane are trained upwards and those from the bottom cane downwards. Shoots from the top canes are held by two pairs of moveable foliage wires, and tucking operations are the same as for VSP. Shoots from the bottom canes are separated from those of the top canes and tucked in front of a moveable foliage wire 2-3 weeks before flowering. The wire is moved downwards near flowering. The shoots are thus positioned downwards. The system can also be adapted to spur pruning, although there is less experimental support and commercial experience. For the moment, this method should be applied after local evaluation.

Benefits

The Scott Henry system has three principal advantages over the VSP trellis. Firstly, the canopy surface area is increased by about 60%, because the canopy now has the combined height of the upward and downward shoots. This alone gives the system a higher increased potential for photosynthesis and yield over the VSP. Secondly, shoot density is about halved, since only shoots from the top canes are trained upwards, with the remainder trained downwards. Hence, the canopy is less dense and fruit exposure is increased. The third benefit is that about half the shoots are devigourated since they are trained downwards.

Comparing Scott Henry with VSP vines pruned to the same node number gives yield increases of up to 30% in New Zealand. Sugar is always increased, by up to 0.7° Brix, and acidity lowered by up to 1.4 g/L. Similar results have been found for Cabernet Franc, Chardonnay, Mueller Thurgau and Sauvignon Blanc. Averaged over these four varieties and over two seasons, the yield increase was 18%, sugar increase was 0.4° Brix and titratable acidity declined by 0.6 g/L. Studies to date have shown no difference in yield or fruit composition between the two tiers. For two years we have had experimental wines made of limited Cabernet Franc trials from Gisborne, and these have been assessed by wine industry judges. Wines from Scott Henry scored significantly higher marks than VSP, with better colour, fruit character on nose and palate, and palate structure. Overall quality rating was 23% higher.

Typically the Scott Henry is an easy system to convert to, especially from VSP. For conversion, head height has to be about 1 m (3 ft). It is easy to harvest with existing machines, and no trouble has been encountered in New Zealand with the downward shoots holding open catching plates of the harvester.

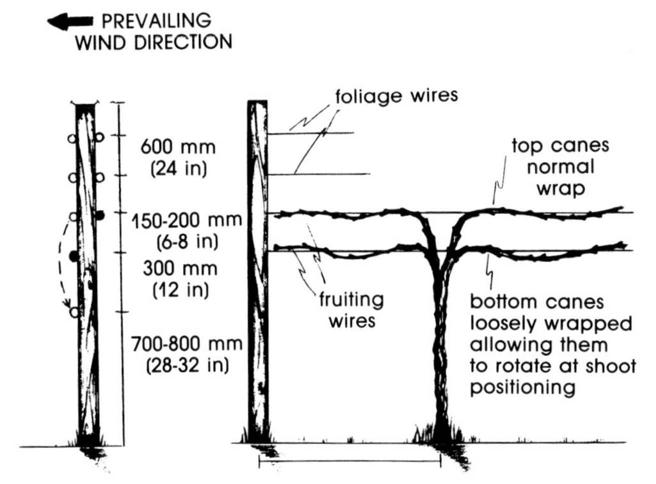

End and side view of cane pruned Scott Henry system. Note top canes can be tightly wrapped but bottom ones should be loosely wrapped. The distance from the soil to bottom fruiting wire should be a minimum of 1000 mm (39 in), to allow for at least 12 internodes per shoot.

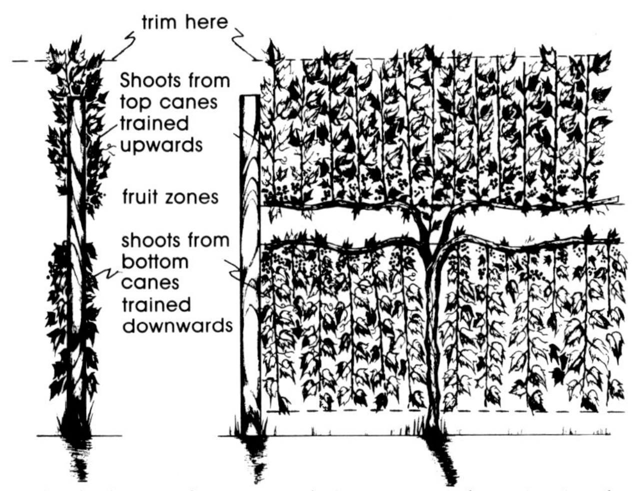

End and side view of Scott Henry during summer to show trimming plane and fruit zone.

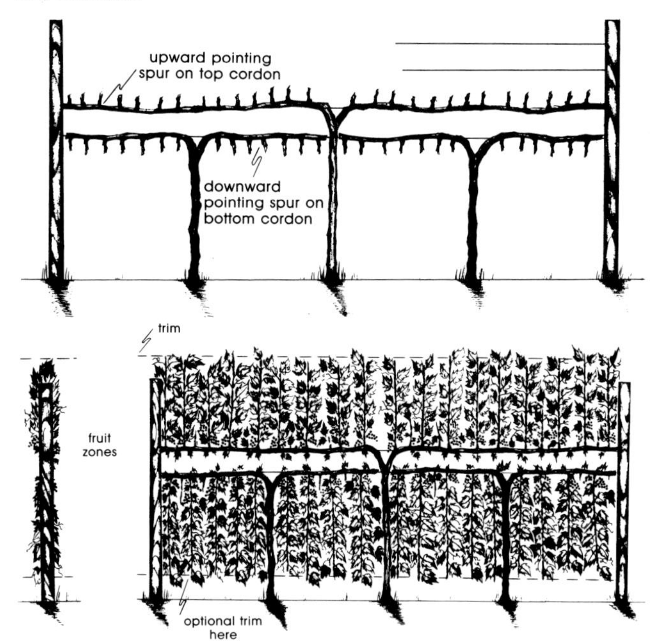

Training vines to spur pruned Scott Henry. One vine is trained to one height only. Note: if there are uneven numbers of vines per panel, put the longest cordons on top. Cordons can be extended past intermediate posts. Mechanization is facilitated if vines are planted near the posts.

A panel of Sauvignon Blanc vines with lower shoots tucked in front of the movable foliage wire. Hawkes Bay, New Zealand. (Photo R.S.)

Larry Morgan moves the wire down to achieve downward shoot positioning on the above vines. (Photo R.S.)

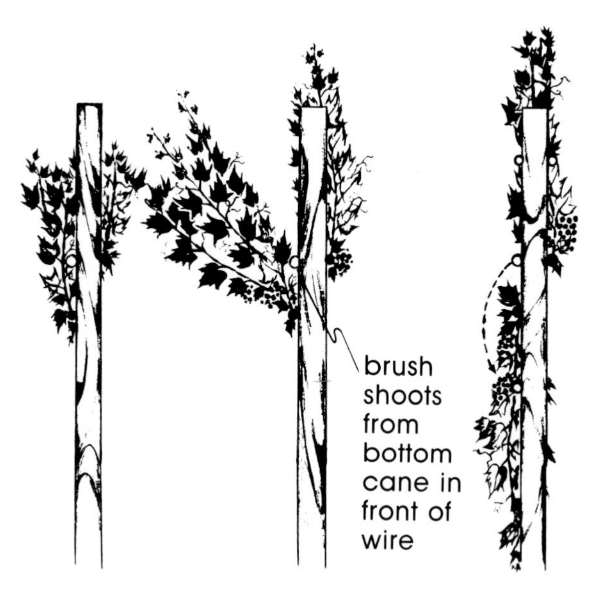

After budburst, shoots grow upwards and out from the two cordons or canes. About three weeks before flowering, brush the shoots from the lower cordon (or cane) in front of the moveable wire. At about flowering, move the wire to the new position which will turn down the shoots (diagram on the right).

Disadvantages

Training shoots downwards is a little more difficult than training them upwards. About 7 hrs/ha (3 hrs/ac) are required to do shoot separation, and then about a further 7 hrs/ha (3 hrs/ac) to turn the shoots downwards. Also the shoot tips of the downward growing shoots may need to be managed because in time these will turn upwards and grow up into the canopy. They should be kept trimmed near the vineyard floor. In dry climates the shoot tips can be allowed to grow on the vineyard floor. Some vines have brittle shoots, eg. Sauvignon Blanc, but this has not prevented the Scott Henry from being used. However, it is critical that the shoots are positioned at the right time.

Management tips

The shoots from the top cane are trained upwards. This requires 2-3 passes and about 14 hrs/ha (6 hrs/ac). Timing is critical for turning the bottom shoots downwards. About 2-3 weeks before flowering, and once the shoots are firmly attached to the cane or spur (when they are not firm they break out easily), shoots from the bottom canes should be tucked in front of the moveable foliage wire. The shoots will begin to pivot downwards under their own weight especially if the fruiting cane is loosely wrapped. It helps to have the bottom wire on the downwind side, because the wind can help tip the shoots downwards. Then, at about flowering, make another pass through the vineyard to move the foliage wire downwards, catching the shoots with it. These downward pointing shoots may need to be trimmed after a week or so.

Similar principles apply to spur pruned vines. Since this system is less well evaluated, caution is suggested with its commercial adoption and trial areas should be established. Two parallel cordons are needed, one for the upward shoots and the other for the downward shoots. It is necessary to train one vine to only one height. Thus one vine is trained to the top cordon, the adjacent vine to the bottom. Spurs point upwards on the top cordon, and downwards on the bottom cordon. This facilitates shoot positioning in both directions. Downward spurs give a shoot growth that is initially downward pointing. For some varieties, pruning to 3 to 4 node spurs may help downward shoot positioning. With long spurs the number of nodes per unit cordon should be kept the same. Ensure that herbicides are sprayed before the bottom shoots are turned down.

Since canopy density is reduced, leaf removal may not be required. At winter pruning, ensure that node density per metre of row is about right so that shoot density will not be too high. When cane pruning, an internode length of about 75 mm (3 in) gives 15 nodes/m (5 nodes/ft). This corresponds to a moderate vigour cane. When spur pruning, have about 7 or 8 two node spurs/m (2.5 spurs/ft). Ensure the bottom fruiting wire is no closer than 1 m (3 ft) to the soil, or there will be insufficient shoot length. Also, the top shoots should extend no more than 300-400 mm (12-16 in) past the top foliage wire.

Since the height of the foliage wall is increased, remember to adjust your spray unit to give full coverage.

Material requirements

Use a 75-100 mm (3-4 in) diameter treated post, 1.8 m (6 ft) out of the ground and spaced 6-8 m (20-26 ft) apart in the row. You will need two fruiting wires and five foliage wires.

Retrofitting

This system, like VSP, is easy to install with existing vineyards. Since the canopy height is about 2 m, the Scott Henry should not be used on rows less than 2 m apart.

Henry Estate Vineyard
Umpqua, Oregon, USA

Development of the Scott Henry Trellis

Significant improvements in both wine and grape yield have been obtained at the Henry Estate Vineyard by the development of the Scott Henry Trellis System. Our vineyard is situated in the Coles Valley floor of the Umpqua Valley. The soil is very rich, composed mostly of a clay loam with soil depths varying between 4.5 - 6 m (15 - 20 ft).

Vine growth and wine history

The first planting of 5 ha (12 ac) was in 1972 and consisted mostly of Pinot Noir and Chardonnay. Trellising was a standard three wire vertical trellis on a 3.6 x 2.1 m (12 x 7 ft) spacing. The plants were headed at the first wire and cane pruned to two canes (typical California system). All vine growth was trained upward. As the vineyard matured, control of growth became a big problem. Wines from early vintages such as 1978, 1979, and 1980 harvests won many medals but less and less of the crop was utilized for these top grade wines. By 1980 (8-year-old vines), a very obvious negative factor from the canopy had been experienced in the fruit and wines. Because of the dense canopies and associated shading, wines were less fruity, had higher pH, and less colour (especially Pinot Noir) than in earlier years. In addition, more fruit was being lost to mildew and bunch rot.

Many standard modifications were tried to improve grape quality, such as shoot thinning, crop level variations, hedging and leaf removal. These modifications only had a slight effect on wine quality. The degradation in fruit quality reached a low in 1982 (10-year-old vines) when a light coloured Pinot Noir was produced with low fruit intensity. Also, all our Chardonnay was lost to bunch rot.

A few of our trials in the vineyard had positive results. First, using four canes instead of two allowed crop level to be increased to offset vine vigour. This reduced bunch rot development. However, increased crop level tended to delay ripening. Also, hedging the vine reduced canopy shading but also resulted in a delay in ripening. Lastly, leaf removal resulted in better wine but required many man hours with its associated increase in cost. What was needed was more leaf area to offset delayed ripening and less leaf density in the fruiting area without incurring a lot of hand labour. This requirement was answered by development of the Scott Henry Trellis System.

Using the system

I prefer 250-300 mm (10-12 in) spacing between the two fruiting wires, while trials in New Zealand utilise 150-200 mm (6-8 in). The wider spacing is preferred to ensure that the window opened up between the two fruiting levels does not close later in the season. This allows the air to move through the window and quickly dry out the fruit if there has been rain or heavy dew.

We have varied crop level in our trials from 5 t/ha (2 t/ac) to 30 t/ha (12 t/ac). Surprisingly, the best wines do not come from the lowest crop levels. At low yields, the vine responds with an increase in the number of laterals. This results in an increase in second crop production, in shade, and a large number of growing tips. High yields will lead to a delay in ripening which is also undesirable. Our bud counts are controlled during pruning and shoot thinning to reach a desired crop level of about 15 t/ha (6 t/ac) for Pinot Noir and 20 t/ha (8 t/ac) for Chardonnay.

Management tips

Division of the canopy occurs about two weeks before bloom (flowering) and extends through bloom. It is important to wait until sufficient growth has occurred to make the movement of the catch wires effective. However, do not wait until the tendrils have toughened which will create shoot damage during separation. Early separation will allow more light into the bunch area which results in better set and better fruitfulness next year. Separation is done by hand with two people (one on each side of the row). The lower growth is separated from the upper growth and wiped downward (often the whole cane turns) and the catch wire is moved and positioned about 300 mm (12 in) below the fruiting wire. Two people should be able to separate about 0.6-0.8 ha/day (1.5-2 ac/day).

Responses to the system

Fungal sprays are more effective because the fruit is exposed. The number of applications of spray may also be reduced. However, the double level of fruit will require a slightly wider spray pattern. Our harvest is done by hand. Pickers are very happy with the system since the fruit is exposed and easy to locate. This results in very little fruit being missed during the harvest process. Prior to the adoption of the system in our vineyard, we could expect about 15% of the Pinot Noir and 25% of the Chardonnay to be lost to bunch rot. These losses have now been essentially eliminated. Fruit ripens 4 to 6 days ahead of a trellis that was hedged but not separated. In addition, the pH of the juice has been reduced which has increased wine quality. Fruit intensity has been improved and a significant increase in colour has been achieved for Pinot Noir. This occurs because of better light penetration into the fruiting zone of the vine and the additional exposed leaf area from the downward canopy.

Utilization of the Scott Henry system will result in a significant increase in crop level for the same total bud count on the vine. This occurs because of better bud fruitfulness since the buds on the replacement canes have been better exposed to sunlight. Modification time and costs to adapt to a Scott Henry system are minimal if a vertical system was previously used. Some additional costs are associated with pruning to four canes instead of two canes and with the separation of the growth between the two fruiting levels. We are now modifying the system using longer canes and with any one vine being trained to just one height.

The Scott Henry Trellis System was developed by a grower trying to solve a canopy problem without incurring extra costs. This system has been largely responsible for bringing our winery back onto the medal winning road. It will not be the answer for everyone. If it is used, a good idea is to conduct some trials in the vineyard before committing significant amounts of acreage to this system. This experience will also enable the grower to try some modifications that may be more suitable for his/her particular site. I would like to thank Richard Smart for his important contribution to the enhancement of this trellis system. If this system does help some growers, my reward is that better wine always makes for a better world.

Scott Henry
Proprietor

Scott Henry and his vineyard.

John Cassegrain and Richard Smart discuss vines trained with one high and one low to establish a spur pruned Scott Henry trellis, Hastings Valley, Australia. (Photo P.M.)

Rachel Smart beside a well tended Merlot vineyard trained to Scott Henry. The vines are covered by netting to protect against birds. Goldwater Estate, New Zealand. (Photo R.S.)

John Cassegrain (Photo P.M.)

Cassegrain Vineyards
Wauchope, NSW

Success with Scott Henry in the Hastings Valley

Cassegrain Vineyards is a producer of premium wines in the Hastings Valley on the East coast of Australia. The climate is characterized by warm summers with high humidity and rainfall, so fungal disease avoidance is a major problem. We first heard of the Scott Henry from our consultant Richard Smart and evaluated it with six trial rows in our Fromenteau vineyard in 1987. The yield increase was a dramatic 70%, so we subsequently converted all of the vineyard. The yield increase has been maintained at 30% and 90% for the following two years. This yield response has been greater than we might have expected but at the same time we have greatly improved our soil management using lime application, slotting and under vine mounding. The vigour has improved and the Scott Henry trellis handles more shoot growth well. Despite the yield increase there has been no negative effects on quality, rather we found pH is slightly improved and acidity higher, a bonus for our wine quality.

We put in new posts to achieve extra height but if we were to repeat the exercise we would attach post extenders to our existing short posts.

During 1987 we were also beginning to decide on a trellis system for our new vineyard developments of Le Clos Sancrox and Le Clos Verdun, totalling 150 ha (375 ac). We decided on the Scott Henry as we wanted to machine harvest and mechanically pre-prune, as well as have a system that will give the best yields and quality for our projected moderate vigour situation. Another advantage of the Scott Henry is that we can in the future change to any other system we want, since we have a head height at about 1100mm.

One vine is trained to the top wire and the adjacent vine to the bottom wire. This means that we can use spur pruning which will help get our shoot spacing right and allow cost savings by mechanically pre-pruning. The trellis cost more to install than the normal Australian trellis, with taller posts and stronger end assemblies. However, we are well pleased with the results to date, particularly with the open canopies which saves us money because of more efficient disease control.

The Hunter Valley is noted for problems with Botrytis bunch rot due to wet conditions at harvest. In the second year of operation of our 'Colline' vineyard there we converted to Scott Henry on 1.6 ha (4 ac) of Shiraz. To develop the cordons we left 45 buds on the Scott Henry and there were about 30 on the vines we did not change. The 1990 vintage was very wet and the results were dramatic. While neighbouring vignerons had to pick Shiraz because of bunch rot, we were able to leave ours on the vine for another two weeks allowing the fruit to build up better maturity. The resulting wines were dramatically improved with more colour, flavour and balance. Also the yield was increased by 100% compared to non-converted vines on the same vineyard. This was due in part to leaving more buds but also due to less bunch rot losses. We were pleased we used Scott Henry and intend to convert all of our vineyards using post extenders in the future.

John Cassegrain
Winemaker

Goldwater Estate
Waiheke Island, New Zealand

Conversion to Scott Henry Training System

Goldwater Estate is a small vineyard dedicated to the production of premium wines — a single blended red Bordeaux style from Cabernet Sauvignon, Merlot and Cabernet Franc, and a Graves style Sauvignon Blanc. The vineyard is situated on an island near Auckland, where the summers are drier and the soils poorer than the nearby mainland. However, the site would be considered to be of moderate to high potential and the control of vineyard vigour can be a problem in seasons of higher than average rainfall. It was for this reason that we considered adopting vine training systems that could utilize this vigour so that shade was reduced and fruit exposure enhanced.

Our initial system was a standard vertical trellis spur pruned to about 50-60 buds per vine, and this was adapted to a lyre or U by dividing the canopy. However, the cordons were not the recommended distance apart at the base. This system proved beneficial in coping with the vigour problem. It gave both increased yield and fruit quality but was unduly labour intensive and difficult to manage, requiring extensive leaf removal around the fruit zone. Thinking that there had to be a better way, we planted a trial of two rows each of Cabernet Sauvignon and Merlot, and trained the vines to the Scott Henry system. The management of this system was compared to the U over a period of four years, and found to require considerably less man hours in both pruning and summer trimming and required little or no leaf plucking. Disease pressure was reduced and fruit quality enhanced, (although at slightly reduced yields). Our vineyard is now entirely converted to the Scott Henry system.

An unexpected bonus with Scott Henry is that with the bottom fruiting wire at 1 m above ground, and with the initial shoot growth vertically upwards, sheep could be left to graze in the vineyard until about flowering. This controlled both undervine weed and watershoot growth. Only one application of herbicide is now required just before the rolling down of the bottom canopy.

Little extra cost is involved in converting from the standard training system to Scott Henry — only two extra wires. The increased fruit quality more than compensates for this.

Kim Goldwater
Owner/Winemaker

Cullen's Willyabrup Wines
Cowaramup, Western Australia

Experiences with the Scott Henry Trellis in Western Australia

In 1988 it was recognized that the yields on Cullen's Vineyard had decreased to uneconomic levels, particularly with Chardonnay and Cabernet Sauvignon. Canopy management, via trellis change, was seen as one way to increase the yield and maintain fruit quality. On the basis of pruning weights, the Scott Henry trellis design was chosen for these two varieties.

The cost for conversion in 1988 from a standard T-trellis was \$2500/ha (\$1000/ac). This included post extensions, and \$1200/ha (\$400/ac) in training costs. With the Scott Henry the most important procedure was brushing down, i.e. separation of shoots on different canopies. Timing of this operation is crucial. In the first growing season, due to our lack of experience with the trellis, we brushed down far too late. This led to difficulty in separating the shoots, as tendrils had bound the canopy and caused excessive shoot breakage. Also, the Chardonnay fruit suffered a little from sunburn due to late positioning.

For Chardonnay, the yield was the same on Scott Henry the year after conversion as it was on the T-trellis (6.3 t/ha, 2.5 t/ac) and fruit composition was similar. For Cabernet Sauvignon, the results were dramatic. Yield increased from 6.4 t/ha (2.5 t/ac) for the T-trellis to 10 t/ha (4 t/ac) for the Scott Henry. There was little difference in the fruit composition, but we preferred wine from the Scott Henry as having more finely structured tannin and blackcurrant flavours. Also, it did not have the herbaceous character and heavy tannin of the standard trellis.

Results from the 1990 and 1991 vintages have shown increased yields for the Chardonnay vines converted to Scott Henry — up to 10 t/ha (4 t/ac). There has been no drop in quality. The Cabernet Sauvignon yields of 1990 and 1991 have remained the same for Scott Henry, with excellent berry characters in the wine. We are proceeding to retrellis the vineyard to Scott Henry after these satisfactory trials.

Vanya Cullen
Winemaker

Kim Goldwater

Vanya Cullen

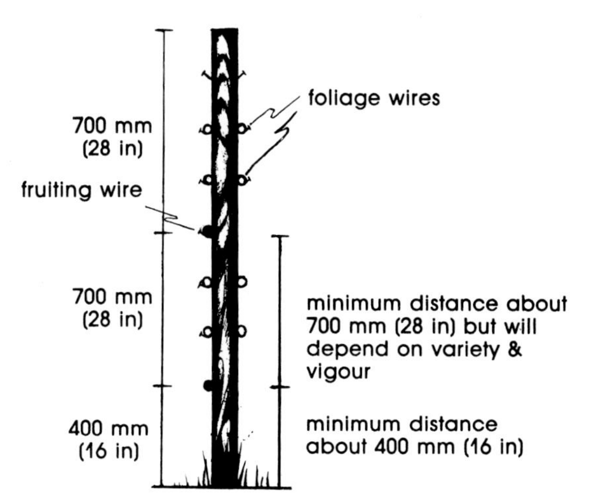

End view of TK2T trellis system post showing dimensions. Foliage wire positions are about 200 mm (8 in) apart. The distance from the soil to the bottom cordon should be set with worker comfort and machine harvesters in mind. It should probably not be less than 300 mm (12 in). The distance between tiers must allow for shoot length of at least 10 nodes, plus about a 100–150 mm (4–6 in) gap.

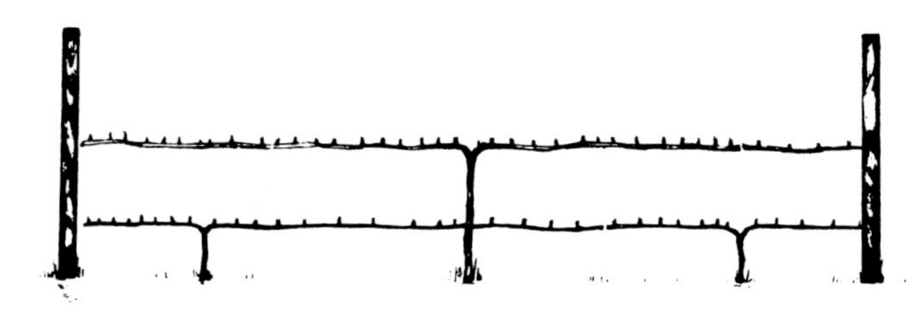

Side view of TK2T showing vine training and spur pruning. Use about 8, two bud spurs per metre of cordon (or about 3 per ft).

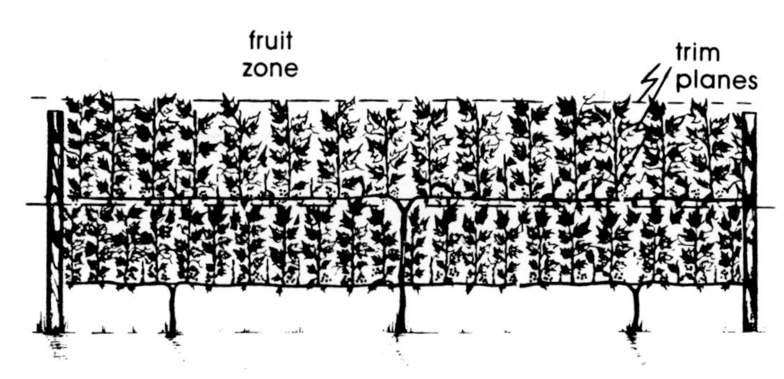

Side view of TK2T during the summer showing fruit zone location and trimming planes.

Te Kauwhata Two Tier (TK2T)

Background

The Te Kauwhata Two Tier system was first installed by Richard Smart at the Te Kauwhata Viticultural Research Station in 1982. He has subsequently seen vines trained to a similar system near Geisenheim in Germany, at Clare in Australia, at Franschoek in South Africa, at Sonoma in the USA, and Hastings and Gisborne in New Zealand. However, none of these vineyards had the same attention given to shoot positioning and trimming as is required for a successful TK2T system.

The TK2T was conceived to be compatible with machines for harvesting and winter pre-pruning, and also it could be retro-fitted to existing vineyards. Commercial adoption has been limited while there remains questions about fruit composition and yield differences between the two tiers. Some ideas to reduce these differences are presented here.

Description

The TK2T is a vertically-divided trellis system, with both tiers of shoots trained upwards. It contrasts with the Scott Henry system which has the bottom tier trained downwards. The bottom tier of the TK2T is about 40 cm (16 in) from the ground, and the top tier about 110 cm (43 in). Although these dimensions can vary, shoots from both tiers are trimmed to about 10-12 nodes length (60-70 cm, 24-28 in) with a gap of about 10 cm (4 in) between tiers. This gap needs to be larger for long-cluster varieties. For each tier there is one fruiting wire and two pairs of moveable foliage wires. Each vine should be trained to one height only, otherwise growth on the bottom tier is depressed. The system is suited to spur pruning because long cordons are required.

Benefits

The TK2T has improved yield and quality over the VSP. This is because the TK2T has a larger surface area and lower density canopy. Yield increases measured in New Zealand have ranged from 26 to 120% depending on variety and year. With large yield increases, fruit maturity is delayed. Low pH also occurs because the shoots are trimmed short. The system can be used in narrow row spacing — say down to 1.8 m (71 in). The system can be machine harvested with existing machines, though the shaking fingers may need to be spread out vertically to accommodate the wider fruit zone. A machine fitted with an appropriate dodger to go around trunks and posts can be used for both summer and winter pruning.

Disadvantages

The principal disadvantage of the TK2T is that the bottom tier has lower yield and delayed fruit maturity compared to the top tier. This is because of partial shading of the bottom tier by the top tier. The maximum maturity difference between tiers we have found is 2.7° Brix — more commonly the figure is about 1.5° Brix. Yields of the bottom tier are about 75% of those of the top. It may be difficult to retrofit a vineyard to TK2T. Commonly head heights are higher than the level of the bottom cordon. This means that vine training is unusual and looks untidy with cordons trained below the head. However, it still seems to work well.

Management tips

The TK2T is an easy system to manage. All shoots are positioned upwards and require only two passes — a total of about 15 hrs/ha (6 hrs/ac). However, trimming takes longer than for VSP and needs to be done carefully between the two tiers. In the first year when cordons are established as canes, it may be necessary to trim the bottom tier shorter than in future years so as to avoid cutting clusters from the top tier. After the first year the vines are spur pruned which raises the top tier fruit zone. A cutter bar with dodger mechanism to avoid trunks and posts will make both winter pre-pruning and summer trimming easy. Since the fruit zone is spread out, mechanical leaf removal may require two passes by a machine. The distance between tiers may need to be altered for different varieties. Allow for a least 10 to 12 nodes in shoot length and a gap of 10 cm (4 in) between the cordon of the top tier and the shoot tips of the bottom foliage.

Maturity differences between the two tiers can be reduced by training one big vine on the top tier, with two small ones below. Also you can prune the top tier more lightly, i.e. 15-20 nodes/m (4.5-6 nodes/ft) and the bottom one more severely, i.e. 10-15 nodes/m (3-4.5 nodes/ft), and also trim the top tier shoots shorter than the bottom tier. Vines on the bottom tier can alternatively be bunch thinned.

Material requirements

The TK2T requires a 1.8 m (6 ft) post out of the ground, two fruiting wires and four pairs of moveable foliage wires.

Retrofitting

The only difficulty with retrofitting the TK2T is that for vines with a high head the new arms for the bottom tier need to be initially trained downwards. While it looks untidy there seems to be no remedy for this, and production is sustainable.

The original Te Kauwhata Two Tier on Traminer at the Viticultural Research Station, Te Kauwhata, New Zealand. (Photo R.S.)

Ross Goodin harvesting TK2T at Te Kauwhata, New Zealand. (Photo R.S.)

Accidentally cutting a vine trunk demonstrates nicely that one vine is trained to one height only. Te Kauwhata, New Zealand. (Photo R.S.)

Martinborough Vineyard,
New Zealand

Experiences with TK2T in Martinborough

The Te Kauwhata Two Tier has been utilized by Dr Jack McCreanor on two, one hectare blocks of Chardonnay and Mueller Thurgau in the Martinborough district of New Zealand.

Establishment is straightforward but with the following considerations. Young plants must be well supported to give straight trunks, perhaps permanent stakes are needed for the high plants which may take longer to establish. Cordons should be lightly wrapped and taped into position as they grow, with future spur positions considered.

Seasonal maintenance is reduced using TK2T, especially if the correct number of foliage wires is used. Two trims per year are required to keep lower shoot growth out of the top tier. The two bunch lines require twice the spray coverage and twice the leaf plucking area but, in general, fruit exposure is better than standard New Zealand trellis or most U frame designs.

The TK2T commits the grower to spur pruning with the design lending itself to mechanisation. Fruit quality would seem to be excellent from three widely varying seasons. Flowering appears to be later when compared with Scott Henry or standard New Zealand trellis. The upper fruit zone ripens earlier and more evenly. Fruit is evenly spread and is easily harvested by hand or machine. The juice has a relatively low pH and acidity allowing the resulting wines to have excellent balance and elegant fruit flavours in styles which cellar well. We have produced wine that has won a gold medal and a trophy for the best wine of the variety at the 1988 Air NZ Wine Awards from the TK2T fruit.

Larry McKenna

Larry McKenna
Oenologist/Managing Director

Cabernet Sauvignon vines trained to TK2T, Sonoma County, California. (Photo R.S.)

Larry McKenna standing in front of Mueller-Thurgau vines trained to TK2T.

Diane Kenworthy and Michael Black

Simi Winery
Healdsburg, California, USA

Experiences with TK2T in Sonoma County

A portion of Simi's 100 acre Alexander Valley vineyard proved to have moderate to high vigour, due to the clay loam soil and favourable exposure. It was decided at pruning in 1984 to retrofit 2 hectares (5 ac) of Cabernet Sauvignon from the California standard non positioned and non divided canopy to a TK2T. The relatively close row spacing (3 m or 10 ft) made the TK2T more attractive than a horizontally divided trellis such as the Geneva Double Curtain.

The trellis follows the design of Richard Smart, although slightly more space between the two tiers allows for the tendency of Cabernet Sauvignon to have longer internodes. Two fruit wires were used, one at 56 cm (22 in) the other at 130 cm (52 in) from the ground. Two sets of moveable shoot positioning wires were added. Alternate vines were established at the lower level and the upper level to avoid apical dominance effects.

The canopy was fully established in the 1985 growing season, although the cordon arms were not completely extended until 1987. Wine has been made in commercial quantity since 1985, and compared to the adjacent block of California standard trellising, which is the same soil type, rootstock and scion selection. The TK2T yields better than the standard. For example, in 1988 the TK2T was 12.5 t/ha (5 t/ac) and the standard 7.5 t/ha (3 t/ac). Wine quality from the TK2T has been very good to excellent, and consistently preferred over the adjacent block. Wine pH is lower from the TK2T by 0.2 units and colour improved.

We have been assured by several companies that we could mechanically harvest the TK2T. Shoot positioning is easily accomplished in one pass. For us hedging is most easily done by hand. The TK2T retrofit required $350/ac for materials and 33 man hours to install. And each year shoot positioning and hand trimming each take about 20 hrs/ac. These additional costs have been well covered by the increased quality and quantity. We currently hand harvest, and our pickers love to harvest the TK2T since fruit access is so easy!

One difficulty encountered is the tendency of the upper tier to ripen earlier than the lower tier, generally by about 0.5° Brix. The lower tier ripening delay was more dramatic in 1988 (1.4° Brix) due to longer internode distance on the lower tier and some sagging of the upper tier, which led to fewer leaves on the bottom shoots after trimming. This delay caused us to harvest the lower tier separately from the upper tier. Also, the entire block ripens later than the adjacent block by up to 3 weeks in normal years. In cooler than normal growing seasons like 1989, the TK2T was not late in comparison to the adjacent block, and both were harvested in the first week of October. We do not therefore believe the ripening delay is a serious drawback in our situation.

In summation, we have been satisfied with our experience with this trellis, both from a wine quality and yield standpoint.

Diane M. Kenworthy Michael Black

Diane Kenworthy, Viticulturist

Michael Black, Vice President/Vineyard Operations

MAF Manutuke Horticultural Research Station
Gisborne, New Zealand

Experience with TK2T at Gisborne

We have had five years experience with the TK2T, both in commercial vineyards and on the Manutuke Research Station. This experience includes the varieties Chardonnay, Mueller Thurgau and Briedecker.

The TK2T gave increased yields of Mueller Thurgau over the four years measured, although sugar was decreased compared to VSP. The respective average figures for VSP and TK2T were yield 22.1 and 27.6 t/ha (8.8 and 11 t/ac) and fruit sugar concentration 18.1 and 17.4° Brix. TK2T reduced percent bunch rot to 22%, versus 32% for VSP.

The Mueller Thurgau vines were planted in 1982 on the Tombleson's vineyard, and we converted some vines to TK2T in the winter of 1985. The TK2T trellis was developed with one fruiting wire at 60 cm (24 in) height and the other at 150 cm (60 in). Four pairs of foliage wires were required, as there were two tiers of growth to train during the season. Vines were trained with cordons to both tiers. During the season of conversion we were able to retain about the same number of nodes per vine as compared to the standard vines (VSP). The grower had trimmed lightly during the previous season, therefore there was a good selection of long canes available for cordon establishment. Due to the good cordon establishment we did not record a loss of yield in the first season and actually the TK2T outyielded the VSP by over 3 t/ha.

The lower maturity of the TK2T fruit was found to be due to delayed maturity of the bottom tier. In 1989, the top tier fruit tested 17.3° Brix and the bottom 16.1° Brix. We also found that the Chardonnay fruit from the top tier at the Manutuke Research Stations was 23.5° Brix and the bottom fruit 21.5° Brix at harvest. The Briedecker at Manutuke also showed a difference between tiers. For all three varieties on TK2T, there has been no significant difference in the pH and titratable acidity between the two tiers. Also, due to the higher levels of shade, we found the bottom tier is less fruitful than the top. For example, in 1989 the Mueller Thurgau top tier produced 21.6 t/ha (8.6 t/ac) versus 15.9 t/ha (6.4 t/ac) for the bottom tier.

In our experience, the management of the TK2T is uncomplicated and it has been easy to instruct staff on the establishment, pruning, tucking and trimming operations. We have two reservations about the system. Firstly, the trimming intervals have to be sufficiently frequent so that shoot growth from the bottom tier does not grow extensively through into the top tier. Secondly, that the pruning of the bottom tier can be tiring, particularly if a pre-pruner is not used. We found that the pruning of the VSP (cane pruned) took around 135 hrs/ha (54 hrs/ac), whereas the TK2T to spur prune took 70 hrs/ha (28 hrs/ac). Shoot positioning of the VSP requires 18 hrs/ha (7 hrs/ac) and the TK2T 14 hrs/ha (6 hrs/ac). This difference is because fewer passes are needed for TK2T since shoots are shorter than for VSP.

Peter Wood

Peter Wood
Viticulture Technician

Peter Wood standing beside the TK2T at Manutuke Research Station, New Zealand. (Photo R.S.)

Fruit exposure improved by conversion to TK2T compared to VSP. Te Kauwhata, New Zealand. (Photo R.S.)

Peter Wood (Photo D.J.)

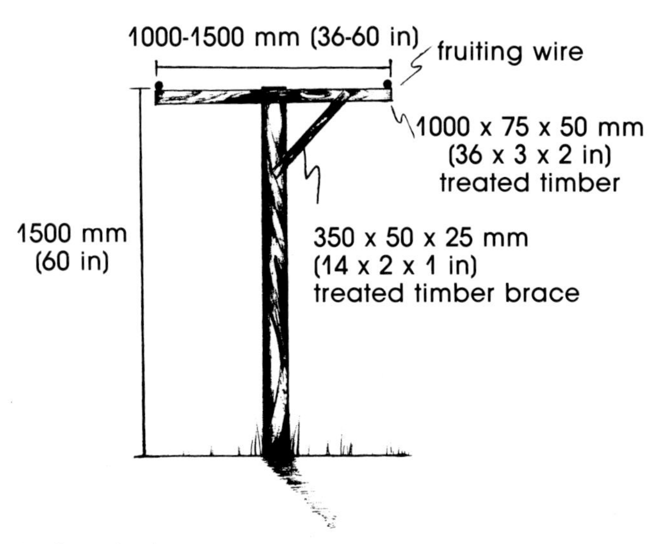

One form of rigid Geneva Double Curtain trellis support using a brace for strength. The GDC should be a minimum of 1 m (3 ft) wide, up to 1.5 m (5 ft) wide.

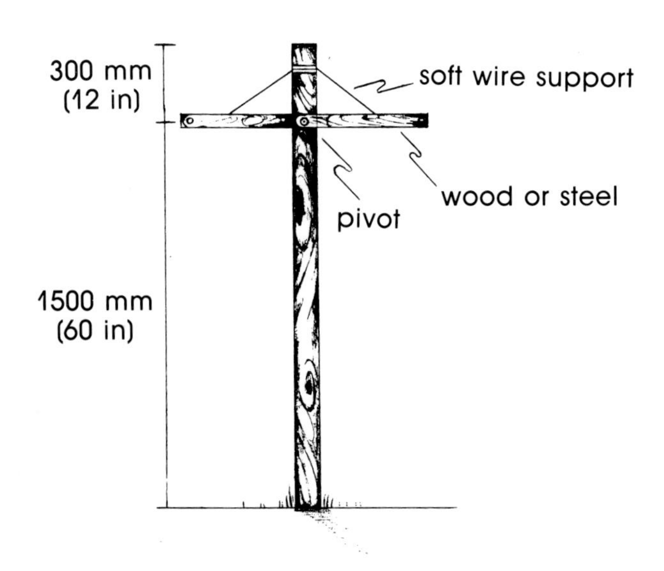

GDC trellis system pivoting in the middle to allow for impact mechanical harvesting.

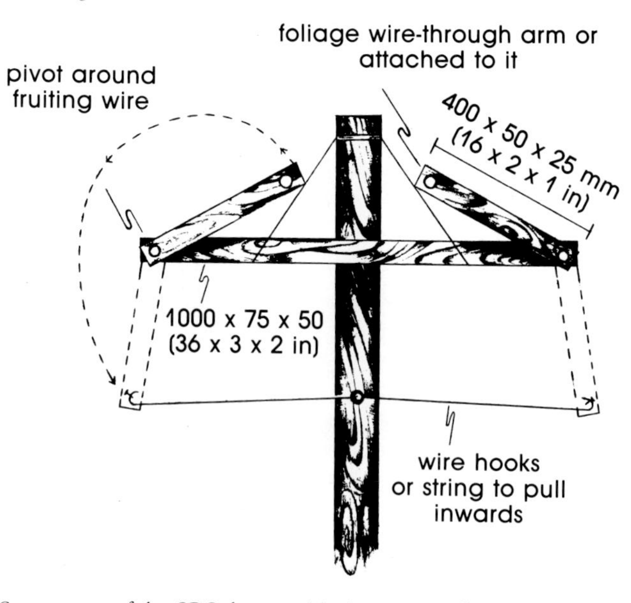

Components of the GDC shoot positioning system. The swinging arm is attached to the fruiting wire.

Geneva Double Curtain (GDC)

Background

In 1966 Professor Nelson Shaulis and colleagues at Cornell University's Geneva Experiment Station first published a description of the Geneva Double Curtain trellis. This paper was a milestone in the development of canopy management theory, elegantly establishing that shade within the canopy was the principal limit to production of quality grapes.

Unfortunately, a stigma was attached to this work because it was done on Concord vines—myths began that these results did not apply to *vinifera* wine grapes, and that *vinifera* varieties could not be shoot positioned downwards. These attitudes remained despite Shaulis and others confirming similar principles with *vinifera* studies. However, Shaulis greatly influenced colleagues from Europe (including Intrieri, Cargnello and Carbonneau) and Smart from Australia. These four people, in particular, have further developed the principles and technology of canopy management for *vinifera* winegrapes.

Description

The GDC is a horizontally divided trellis with shoots trained downwards. The curtains are supported at the top and are free hanging. They are at least 1.0 m (3 ft) apart, and shoots are about 20 nodes (1 m, 3 ft) or longer. The system was specifically designed to improve yield and fruit composition, and its development was parallel to that of the early machines for harvesting and pruning. The trellis was designed to incorporate suitability for mechanisation. The system is normally spur pruned to downward spurs, often greater than two nodes in length. This facilitates downward shoot positioning.

Benefits

Shaulis showed in early studies that two factors were very important to increased yield of mature ripe grapes. Both factors could be included under a general heading of avoiding shade. Experiments with row spacing showed the advantage of increasing effective canopy length (or surface area) per hectare. The second factor involved the position of the renewal zone (i.e. the zone of the base of shoots). Yield and fruit composition were increased if the renewal zone was exposed rather than shaded. Exposure was optimal if the renewal zone was at the top of the canopy, i.e. the shoots were encouraged to grow downwards. So the GDC idea was born, and was made compatible with local growers vineyard machinery in New York. It was easier to divide canopies on wide rows than replant vineyards to narrow rows and buy specialized machinery.

The GDC system is suitable for moderate to high vigour vineyards. Downward shoot positioning causes desirable shoot devigouration. GDC can convert dense, shaded canopies to low density ones with more than 50% yield increases obtained compared to VSP. Fruit exposure is dramatically increased and notable gains in fruit composition and wine quality have been recorded. In particular, pH is lowered and colour and phenolics increased.

The system is very easy to mechanically pre-prune, and cut canes fall freely to the ground. Also, machine harvesting is possible with a range of machines including slappers and vertical impactors. Modification to the trellis supports are required to accommodate different machines. Shoot positioning can be done by hand, by machines available in New York State, or with a moveable wire on a swing arm. This latter idea was developed at Roseworthy College in 1976 by Richard Smart.

Disadvantages

Difficulty with the need to vertically shoot position downwards is the major problem with GDC. Some growers install the system, do not shoot position and wonder why yields and quality do not improve! Most *vinifera* varieties have upright shoot growth, so a few tricks have to be learned to train them downwards. Another problem is that fruit exposure can be excessive with GDC which, for example, can give too high phenol levels, particularly in hot climates. Again, steps can be taken to reduce fruit exposure to acceptable levels. Also, some varieties can produce many fruitful watershoots on the well exposed cordon. These often need thinning to prevent excessive yield.

Management tips

Shoot positioning is facilitated by pruning to downward pointing spurs, formed by rotating the cane and shoots in the first year of cordon formation. Shoot positioning is easier if these spurs are 4 to 5 nodes long rather than the normal two nodes, but still remember to have only about 15 nodes per metre of cordon. Begin positioning the shoots outwards about three weeks before flowering, and turn them downwards about the beginning of flowering, during a second pass. The total shoot positioning time is 15-20 hrs/ha (6-8 hrs/ac). If watershoots and all inward growing shoots are removed from the cordon early in the season, then vigorous varieties like Chenin Blanc will literally shoot position themselves. It is imperative to keep the gap between curtains open so sunlight can penetrate. The most common mistake with GDC is that shoot positioning is not done properly. **If it does not look right, it will not work right!** Over exposure of fruit can be reduced by pruning to longer spurs, so that the fruit hangs below the cordon and is on the side of the vertical canopy wall. Secondly, some current season shoots can be trained along the cordon and taped in place — this will provide an umbrella of leaves to protect the fruit. Labour costs of this last operation will be about 25 hrs/ha (10 hrs/ac).

When young vines are trained, loop the arms over a central wire and run these out horizontally. This will avoid trunk splitting and improve the ease of machine harvesting and pruning. One vine can be trained to both sides of the GDC or alternatively one vine to each side, as was originally described for GDC. This latter approach was to ease the collection of pruning weights per vine, and is not an essential design feature as is commonly thought.

Materials

Intermediate posts should be 1.5-1.8 m (5-6 ft) high. The GDC requires two fruiting wires and if used, two moveable foliage vines for each row. One wire down the centre of the row should be used when the vines are trained to form arms. This wire could subsequently become a foliage wire. The GDC cross arm can be made from 75 x 50 mm (3 x 2 in) treated timber, and the swinging arms for the moveable foliage wire from 40 x 25 mm (1.5 x 2 in) timber. The collapsible form of the GDC suitable for harvesting with a slapper type machine uses 1000 x 50 x 50 mm (40 x 2 x 2 in) tanalised timber supports, two to each post.

Retrofitting

The GDC system is easy to retrofit. Normally this requires long canes, as they must go up from the old head or cordon at say 1 m (3 ft) to 1.5 m (5 ft), out each side by 0.75 m (2.5 ft) and then along a half vine space (say 1 m, 3 ft). The total length of required cane is therefore about 2.25 m (7.5 ft) and four of these are required per vine. (This assumes one vine will be trained to both sides.) At pruning the first year, wrap canes loosely so that downward pointing shoots are produced in the first growing season by downwards shoot positioning. This then creates downward pointing spurs.

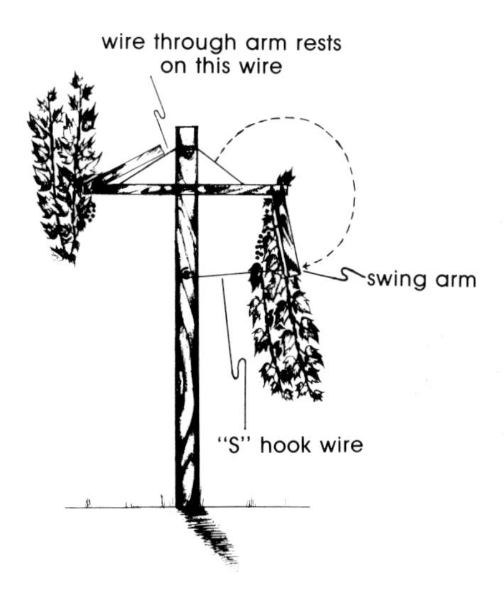

End section to show how the movable foliage wire on the swing arm is pulled down and secured at about flowering to achieve downward shoot positioning. Left side before positioning, right side after.

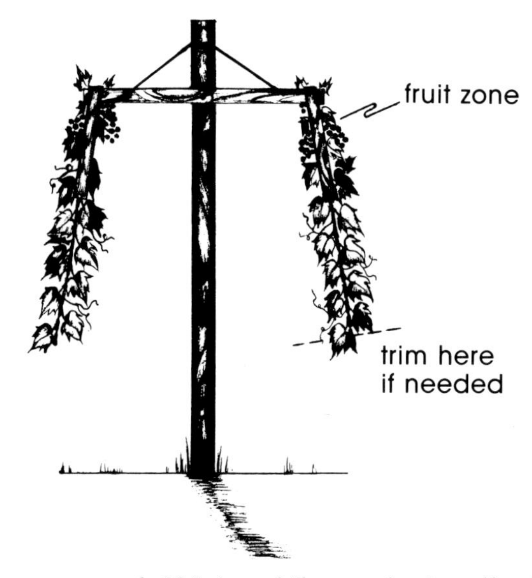

End section appearance of a GDC vineyard. Shoots can be trimmed but this is generally not needed.

Vine training for the GDC. Note how arms are trained around the central wire to give flexibility. Vines can be trained to one or both sides, but cordons must fill the wire. Prune to downward pointing spurs, from 2 nodes to 6 nodes long.

GDC training for traminer at Te Kauwhata, New Zealand. Note high fruit exposure. (Photo R.S.)

Richard Smart standing in a Concord vineyard trained to GDC, Hector, New York, 1974.

Larry Morgan (Photo D.J.)

MAF Hawkes Bay Agricultural Research Centre
Hastings, New Zealand

GDC Experiences in Hawkes Bay

In Winter 1988, Richard Smart and I had discussions with a local grape grower, Nigel Read, who wanted to convert his high wire Sylvoz to a system which gave similar yields but improved fruit composition (i.e. lower rot, and earlier ripening). One of the systems we decided to evaluate was the Geneva Double Curtain, with early summer shoot positioning being an essential feature.

The existing high cordon was given a close machine trim, then hand pruned to leave around 100 nodes per vine. After pruning, the canes were unwrapped off the high wire, and new fruiting wires installed using a fence batten support system, hinged at its base to allow for machine harvesting. Alternate vines were then wrapped onto left and right support wires, leaving some gaps in the cordon to be filled in winter 1989. Two central wires were put in place to use for early summer shoot positioning — these were pulled out over the cordon and hooked onto a nail at flowering time. Total conversion time from the initial machine trim to wrapping the cordons onto the wires and shoot positioning was about 120 hrs/ha (48 hrs/ac).

In the second winter, once both cordons were fully established, winter pruning became a straightforward and quick operation. A tractor mounted trimmer was used to pre-prune the top and side of the cordon after the shoot positioning wire was dropped. This operation was followed by a quick hand prune to remove any long and inward growing spurs. Total winter pruning time was therefore only 10 hrs/ha (4 hrs/ac) which included repositioning the moveable foliage wire. Summer shoot positioning involves lifting the shoot positioning wire out and over the cordon and hooking it under a nail on the batten. It was done around flowering time and took about 10 hrs/ha (4 hrs/ac). The operation is easier with two people.

First harvest off the GDC system was in February 1989. Yield from the GDC in 1989 was equivalent to 26 t/ha (10.4 t/ac) compared with 32 t/ha (12.8 t/ac) on the grower's high wire Sylvoz. Slightly higher yields were expected from the Sylvoz since more nodes were retained. Analysis of the fruit at the winery in 1989 showed the GDC to be 19.1° Brix and titratable acidity of 7.1 g/l while the Sylvoz was 16.0° Brix and 8.1 g/l. This difference was to be expected since the fruit on the GDC was well exposed whereas that on the Sylvoz was shaded. Machine harvesting of the GDC proved to be no problem, as the foliage curtains closed up as the slapper harvester passed over them. A small amount of batten breakage occurred, but was more from inferior construction material (which was on hand at the time of conversion) than a fault of the system.

Wine was made in small scale at the Te Kauwhata Research Station from both the Sylvoz and the GDC. At a December 1989 tasting by industry judges, wine from the GDC was clearly superior to the Sylvoz with 14.6 points compared to 12.2 out of 20.

The 1990 harvest was 41 t/ha (16 t/ac) off the GDC and 28 t/ha(11 t/ac) off the Sylvoz. Despite this 46% increase in yield over the Sylvoz, the GDC fruit maturity was decreased only 1.1° Brix with similar acidity and pH. This large yield increase easily covers the cost of trellis conversion and annual costs of shoot positioning.

Larry Morgan
Viticulture Technician

Villa Maria Estate Ltd
Auckland, New Zealand

Retrofitting a GDC

A Gisborne grapegrower (Donald Gordon) adapted the GDC so as to machine harvest with a standard slapper type machine, which is most commonly used in New Zealand. Our experiences with this New Zealand adaptation of the GDC have been very favourable with significant improvements in yield and quality for the varieties Traminer and Chardonnay over vines on standard trellis (VSP).

The system can be retrofitted from standard trellis at 3 m (10 ft) row spacing and allows the use of normal vineyard tractors. The supports for the two curtains are constructed out of 1 m long 50 x 50 mm (40 x 2 x 2 in) treated timber battens, hinged to a position 0.8 m (32 in) above the ground using a loop of mild steel 4 mm wire and staple. This hingeing allows the wooden cordon supports to be pushed up against the post as the machine passes and this is the key for machine harvesting with slappers. There are two supports per intermediate post each with a cordon wire attached to the top end and inside of the support. The two supports are guyed together across the top of the post, using either 4 mm mild steel wire, or 5 mm diameter nylon boat rope. The supports are held approximately 1 m (3 ft) apart at the top by the weight of the foliage and fruit. The moveable foliage wires are positioned at the top and inside of each support, next to the cordon wire after pruning, and are moved to position the shoots downwards immediately after flowering. They are held in place by a small wire clip and/or staple attached to each wooden support. The extra cost of establishment over the standard trellis is approximately $500/ha ($200/ac) inclusive of labour and materials. This investment is easily paid back in one year.

We decided to train the vines with only one curtain per vine. Vines are alternately trained down the row, from one curtain to the other, each vine with a cordon stretching over two vine spacings. This can help stop splitting of the trunk or cordon by the machine harvester. When retrofitting vertical trellised systems, the conversion is achieved in one or two growing seasons without any loss in crop.

Once established the GDC is spur pruned to outward and downward, two or longer node spurs. The system is easily pre-pruned using either saws or a reciprocating knife cutter bar. However, because the nodes are very fruitful, node numbers need to be regulated by follow-up hand pruning. We leave one node per 8 cm (3 in) cordon in our moderate vigour Chardonnay vineyard. To ensure the canopy remains divided throughout the season, disbudding of the trunk is required up until flowering.

Shoot positioning takes two steps, especially when dealing with an established cordon. The first step is to bring the moveable foliage wire out to position the shoots outwards from the cordon in a horizontal plane. Then, 3 to 7 days later the wire can be moved downwards to position the shoots vertically downwards, and clip the movable wire to the support. The two step procedure prevents excessive shoot breakage, which can happen on some erect growing *vinifera* varieties if positioning is done too early or too late. Some summer trimming is often required on the shoot tips and above the canopy. Machine harvesting simply requires opening the picking head out to give the maximum gap and ensuring the picking fingers are suitably positioned. The hinged supports are pushed up by the guards at the front of the harvester, along with the foliage and fruit. Avoid harvesters with a narrow throat.

Three years of commercial trials have convinced us of the benefit of the GDC for Chardonnay at Gisborne. Average yield was increased over 70% for the same average sugar level. Also, pH was lower by 0.06, and titratable acidity by 0.9 g/L.

Grape composition is slightly improved, except where the GDC is excessively cropped. This occurred in one year with sugar levels 1.2° Brix lower than VSP when yield was almost doubled. Improved sun exposure of the fruit increases varietal character in the wine. For Traminer, GDC fruit is spicy and gingery, with more deeply coloured berries. Our best quality Traminer wines have come from vines trained on GDC—these wines are national wine show winners.

Steve Smith
Viticulturist

Steve Smith in a retrofitted GDC vineyard, Gisborne, New Zealand. (Photo R.S.)

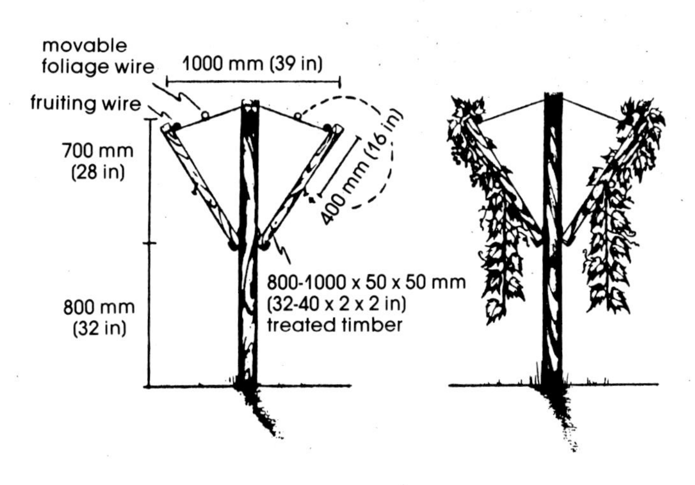

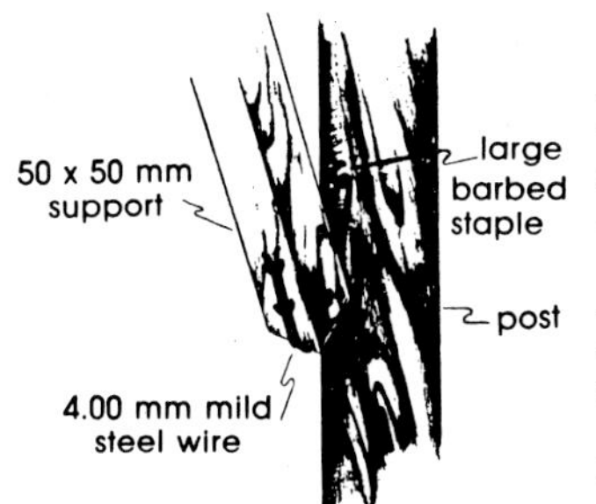

Above left: End view of flexible GDC support designed for mechanical harvesting as described by Steve Smith.

Above right: Shoot positioning is achieved with a movable wire attached to the wooden batten.

Left: Details of construction of the cradle support for wooden arms.

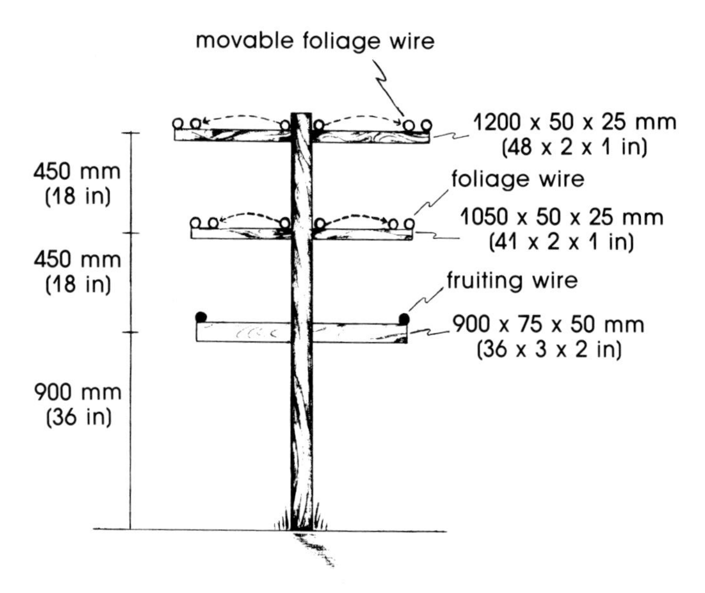

Details of trellis support system for the U using cross-arms on a central post. Minimum recommended distance apart at the base is 900 mm (36 in).

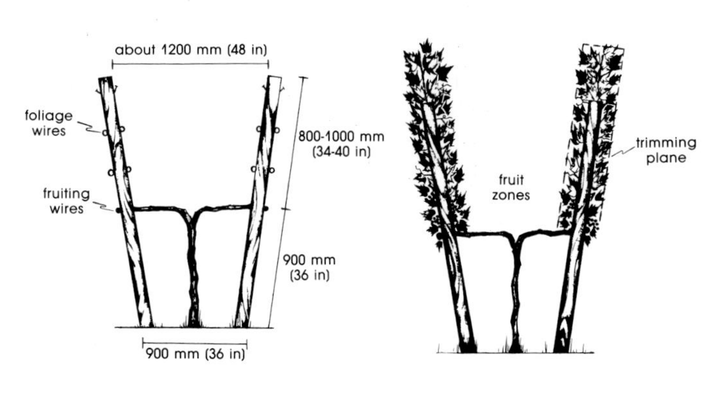

A U trellis system using two inclined posts, normally treated softwood, 100–125 mm (4 to 5 in), 2.4 m (8 ft) long, embedded 0.6 m (2 ft) in ground.

End view of vines trained to the U system in summer, showing trimming planes and fruit zone. Note how trimming the inside of the system is facilitated by using two inclined posts.

Vine training to the U system showing spur pruning, although cane pruning can also be used.

The U or Lyre System

Background

The U or lyre system was popularized by Dr Alain Carbonneau of INRA Station de Recherche de Viticulture in Bordeaux. First results of the system were published in 1978. The U trellis has been evaluated commercially in several countries, and with the availability of a specifically designed machine harvester in 1989 this adoption will increase. The U System is sometimes incorrectly termed 'quadrilateral cordon' in California.

Description

The U system is a horizontally divided trellis with shoots trained upwards. The two curtains should be at least 900 mm (36 in) apart at the base and the foliage walls are inclined slightly outwards. Shoots are trimmed to about 15-20 nodes. The system can be either spur pruned or cane pruned.

Benefits

Because it is a divided canopy, the U has an increased surface area and typically an open canopy, which lead to yield and quality improvements. Yield responses after conversion of 50 percent or more are common. The fruit zone is at the base of the canopy, and so the fruit is not exposed as much as for the GDC. Shoot positioning is easy to achieve. Carbonneau reports significant improvement in red wine quality, especially for soils of higher potential. The U system can be readily machine pre-pruned, and a French harvester that uses an impacter head has been developed.

Disadvantages

The U system will not work properly unless the centre is kept open. Provided there are sufficient shoot positioning wires, and these are moved at the correct times, then shoot training is no problem. However, for high vigour vineyards lateral growth to the centre is difficult to contain. We suggest that the U system is not used on high vigour sites unless some devigoration method is used, e.g. grassing down or vine removal to create big vines etc.

Management tips

Summer trimming of foliage walls is easier if inclined posts are used, rather than a centre post and crossbar. This is especially important if excessive lateral growth is anticipated. Two pairs of foliage wires are required for each side. The outside wire may be fixed, but the inside one should be movable. Do not allow shoots to grow out from the top of the trellis and shade the fruit zone. They should be kept trimmed. Minimum row spacing is normally 3 m (10 ft).

Material requirements

Posts of 1.8 m (70 in) out of the ground are normally used. Cross pieces of wood can be 75 x 50 mm (3 x 2 in) for the fruiting wire support and 50 x 25 mm (2 x 1 in) for foliage wire support. Alternatively, the system can be built with two inclined posts, thus avoiding cross pieces. There are two fruiting wires and eight foliage wires per row of vines.

Retrofitting

This is an easy system to retrofit if the existing vineyard has a crown height of about 800-1000 mm (32-39 in) and row spacing is about 3 m (10 ft). If existing poles are sufficiently tall they can have crossbars added. For situations where problems with lateral growth are anticipated it is better to use inclined posts. Normally vines are cut back and four new canes laid, two to each side. Disbud these canes out to the fruiting wires. For young vineyards, it is often possible to remove cordons from the existing fruiting wire and use these. It is best to alternate vines down the row to each side, pushing the trunk over towards the fruiting wire. This means that existing cordons can be retained, and the fruiting wire gaps can be filled by canes from the end of the cordons in the first year of conversion. Remember, do not trim hard in the year before conversion so there is sufficient cane length to fill the wire. Our experience is that the cost of material and labour usually will be recovered from the extra crop in the first year after conversion.

Kumeu River Wines Ltd
Kumeu, New Zealand

Experiences with the U System in Auckland

We first installed the U or Lyre trellis system in our Kumeu Vineyards in 1983, during the second year of growth of young Cabernet Sauvignon and Chardonnay vines. This was followed a year later with Sauvignon Blanc, Cabernet Franc and more Chardonnay. The vineyards with this system now total 9 ha (23 ac).

The vine rows are 3.3 m (10 ft) apart, with 1.8 m (6 ft) between the vines. We used inclined posts to set up our U system. Each row is divided into two halves which are separated by 700 mm (28 in) at cordon level (900 mm, 36 in high) and 1200 mm (48 in) at the top of the posts (1.8 m, 72 in high). Were we to retrofit more vineyards we would increase these distances by 40% to allow for a larger distance between canopies. This arrangement is achieved using 2.7 m x 100 mm (9 ft x 4 in) posts driven into the ground at the appropriate angle and distance from the centre line of each vine row. There is one fixed cordon wire for each half of the trellis and four moveable foliage wires for training the shoots in a vertical manner. The use of angled posts instead of posts and cross-arms has allowed us to adapt a trimming machine for clearing the inside faces of the trellis. This has proven to be essential to the successful management of the system, as the correct degree of illumination of all parts of the canopy is only attained with the central gap free of excess vegetation.

The results to date from this trellising system have been extremely good. The Cabernet Sauvignon and Cabernet Franc are notable for their vibrant colours, fruity aromas and smooth tasting phenolics resulting in very good red wine styles. The Chardonnay and Sauvignon Blanc produce wines that are now very highly regarded both in New Zealand and internationally.

I believe that the quality of these wines is primarily due to superior ripening conditions afforded by the U system. We have consistently noted wines with richer colours and much riper aromas and flavours than those from conventional systems and the grapes seem to achieve real fruit maturity earlier despite a reasonably large crop load.

Michael Brajkovich

Michael Brajkovich, MW
Winemaker

Michael Brajkovich, MW.

Dr Alain Carbonneau of Bordeaux standing beside a U system in Dave Adelsheim's vineyard, Orgeon. (Photo R.S.)

Trimming machine developed at Kumeu River Wines using three cutter bars to simultaneously trim the top, outside and inside walls of the U system. (Photo B.W.)

Kate Gibbs in the centre of the U system at Kumeu River wines, Auckland, New Zealand. (Photo B.W.)

Traminer vines trained to U system using cross-arms at Te Kauwhata, New Zealand. (Photo R.S.)

A U system badly in need of shoot positioning. Australia. (Photo R.S.)

A retrofitted U system which is too narrow at the base causing shade and restricting the yield and quality potential. New Zealand. (Photo R.S.)

De Celles Vineyard
Napa, California, USA

Experiences with the U System in the Napa Valley

De Celles Vineyard is located in the cool, lower end of the Napa Valley in the Oak Knoll area. The soil is deep so irrigation is not necessary. Dr Richard Smart evaluated our vineyard in the summer of 1988, and suggested a change to the Lyre, or U trellis system. The task of conversion was completed the following winter.

The first task was to pre-prune the Chardonnay vines which simplified the operations to follow. Next, the existing two wire system was removed. The vines were in-grown to the wires in many places, resulting in a painstaking project. But removing the wires and canes made it much easier to later manoeuvre the post driver. The Lyre trellis system was then constructed. Unfortunately, the vineyard had been heavily trimmed the previous growing season, and the canes were too short to fully form as quadrilateral cordons. At this point we decided to rotate each vine 90 degrees, tying the old cordons to the new fruiting wires, and cutting off the excess where they extended beyond the wire. The success surprised even the experts. It was possible to fill the fruiting wire using canes off the old cordons even though the canes were relatively short. We almost doubled bud number per vine.

We were excited with the crop showing on the vines, but 1989 was a difficult harvest in the Napa. The heavy rains which fell in September as picking crews stood by helplessly were followed by three days of high temperatures. Bunch rot began to spread rapidly through the white grape crops. Some vineyards were left unpicked. Due to the ease with which Lyre-trellised vines can dry following rain, we were able to harvest all our crop. Since the grapes are so accessible, each harvest worker was able to pick a hefty weight in a single day! Despite the fact that the new trellis system almost doubled the yield in our vineyard, the fruit maturity remained on a par with adjacent vineyards still using standard trellis systems. The cost of materials and labour to convert was a steep $12,500/ha ($5000/ac), but the increase in yield more than paid for this expense in the first year.

Joe De Celles — Mary Lee De Celles

Joe and Mary Lee De Celles standing inside their U system.

'he Sylvoz System

ackground

he Sylvoz system is popular in Italy, and has recently been used Australian and New Zealand vineyards in an effort to reduce roduction costs. It is well suited to machine harvesting and runing. The system is often termed 'hanging cane' in Australia.

escription

'he traditional Sylvoz is a non divided canopy, with a high ordon. Pruning is normally to vertical canes which are tied own or free hanging. Shoots from these canes tend to grow ownwards under their own weight as they are unsupported. Iowever, some shoots fall along the top of the canopy and shade he fruit zone. There are several variations of the Sylvoz. The 'ines can also be developed with a mid height cordon, with ruiting canes tied below, and replacement canes growing up-vards where they are well exposed. A second variation is to have high cordon but to add two foliage wires which are moved lownwards to provide vertical downwards shoot positioning. Trimming is carried out in summer above the cordon to avoid hading.

Benefits

The main benefit of the Sylvoz is that it saves labour costs in winter pruning. It is easy to machine pre-prune and, even with nanual follow up, pruning time will be less than 25 hrs/ha (10 nrs/ac). Prunings fall freely to the ground.

Disadvantages

The Sylvoz does not give high yields unless pruned very lightly. Trials in New Zealand with Sauvignon Blanc, Cabernet Franc and Mueller Thurgau show similar or reduced yields compared to VSP trellis with the same bud number. Bunch rot on the Sylvoz was always increased. In recent trials, shoot positioning and trimming has been introduced to improve exposure of the bunch zone to avoid bunch rot. This should lead to higher fruitfulness and yield. Unless pruning to downward spurs is carried out, it is difficult to downwards position shoots in summer and canes in winter. Winter pruning of Sauvignon Blanc for example with the mid height Sylvoz leads to some cane breakage as they are pulled down.

Management tips

Prune to downward pointing spurs to facilitate shoot positioning. For high yields, we suggest pruning to five node spurs each 200 mm (8 in). Do not allow foliage to cover the cordon as this leads to reduced yield and increased bunch rot. Shoot positioning, leaf removal and trimming can help avoid this.

Material requirements

The Sylvoz is a simple trellis to erect. A post 1.8 m (6 ft) out of the ground is required. Only one fruiting wire is needed, and if desired, two foliage wires can be added. The mid height Sylvoz requires one fruiting wire with one or two pairs of foliage wires above, and one or two wires below for attachment of canes.

Retrofitting

The Sylvoz is easy to retrofit. Remember to trim lightly the season before so that there is sufficient cane length to fill the wire with new cordons.

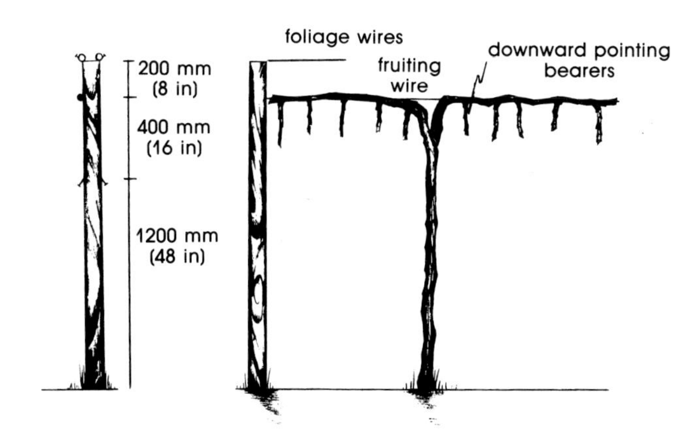

End and side views of vines trained to the high wire Sylvoz. Vines can be pruned to bearers 5–10 nodes long.

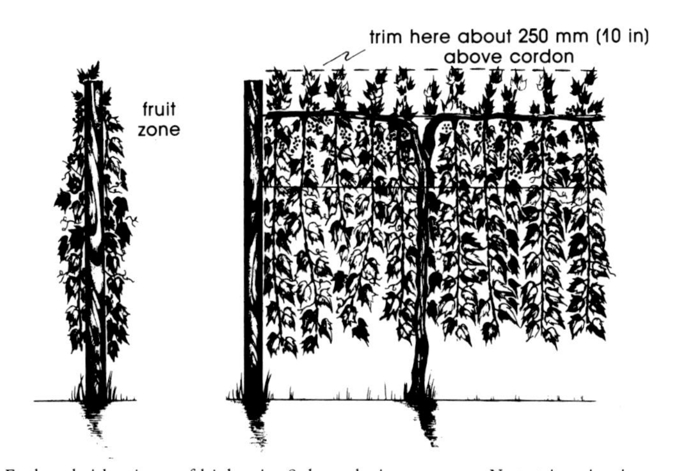

End and side views of high wire Sylvoz during summer. Note trimming is done about 250 mm above the cordon, and movable foliage wires are brought down at about flowering.

High wire Sylvoz at budburst. Manutuke, New Zealand. (Photo R.S.)

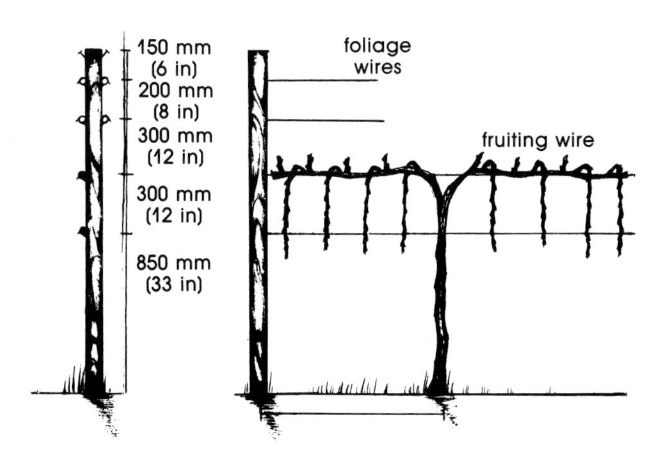

Side view, mid height Sylvoz, pruned to 6–10 node canes and 2 node spurs on top.

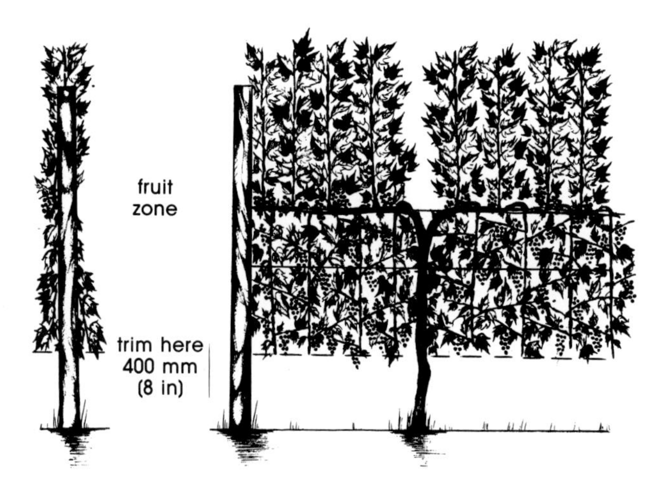

Side view of mid height Sylvoz during summer. Shoots from spurs are trained upwards.

The McDonald Wine
Montana Wines L
Taradale, New Zeala

Developments in Sylvoz Training in Hawkes Bay

Mid height Sylvoz is a training system which has been trialle and developed in Hawkes Bay since 1980. Fruiting canes are ti downwards from a mid height cordon, on which replaceme canes are produced.

The preliminary trials from 1980-1983 were conducted c Mark Read's property and gave the following results on Mull Thurgau. Sylvoz trained vines pruned to about 70 nodes we compared with vertical shoot positioned vines pruned to 3 buds. The average yield for three years was 29 t/ha (11 t/ac) fo Sylvoz and 16 t/ha (6 t/ac) for VSP. Fruit from the Sylvoz had low sugar content by 0.6° Brix. These responses are expected fro a difference in bud number laid down at winter pruning. Fro this period onwards the system was trialled on a number c properties and varieties. Success was achieved with Sauvigno Blanc, Chardonnay, Cabernet Sauvignon and Riesling.

More extensive trial work has been carried out in co-operatio with Campbell Agnew over the 1988 and 1989 seasons. Th system has been modified to intercept more light and conse quently increase yield and quality. This was achieved by attach ing fruiting canes to a 460 mm (18 in) wide tee, located 600 mr (24 in) from the ground. This trial was conducted on Muelle Thurgau vines and has been evaluated on vineyard block scale

Early indications from the first two harvests are that modify ing the Sylvoz gave a yield increase of 20%. The yield potentia of Mueller Thurgau trained to the modified Sylvoz is about 3 t/ha (14 t/ac), about double the district average of 17 t/ha (7 t/ac)

The advantage of the modified Sylvoz training system in term of yield and fruit quality outweigh the minor disadvantages These disadvantages are a risk of delayed maturity if the vines ar overcropped and low bunches which can be difficult to pick with some harvesters.

The advantages of the modified Sylvoz are that fruitful canes can be easily selected, and the increase in canopy surface area gives more light interception and the capacity to carry a larger crop to maturity. Further, the canopy is more open, resulting in better spray penetration and more air movement about the bunches, hence better disease prevention. The grower can increase bud numbers to balance the crop potential of the vine and minimise congestion. There is a significant increase in yield, up to 100%, as a result of the large increase in the number of well spaced bunches.

G. J. Wood

Gary Wood
Vineyards Supervisor

Budburst taking place for mid height Sylvoz. (Photo G.W.)

Gary Wood

inimal Pruning

ackground
n the mid 1970s commercial adoption of mechanical pruning egan in Australia in an attempt to reduce production costs. ver the past two decades CSIRO researchers have been udying minimal pruning, whereby virtually little or no winter runing is done. This practice is now widely used in Australia, ncluding vineyards in hot, irrigated regions as well as cooler, remium quality regions. It is likely that this practice will find ore application in hotter rather than cooler climates. The ystem was introduced to commercial Auckland vineyards in 985 by Allan Clarke of MAF.

escription
he system has been termed **MPCT** (minimal pruned cordon rained) and vines are trained to a cordon on a high single wire. here is little or no winter pruning but during summer shoots re trimmed to keep shoots and fruit off the ground. Shoot rowth direction is uncontrolled. Because of large node numers retained per vine, shoot growth is devigorated and there can e as few as 5 nodes per shoot. The vine develops an extensive ramework of old wood, some of which is removed annually vith mechanical harvesting.

enefits
IPCT vines are easy to machine harvest, and so total labour nput for the system is very low. MPCT is a means of controlling xcessive shoot vigour, as shoot number per vine is higher than ormal. The canopy quickly develops an increased leaf area and urface area in spring, which leads to a higher yield potential. In ome circumstances fruit exposure can be increased and canopy ensity reduced. Averaged over several years, yield is slightly arger than normal pruned vines, though large crops can result n the first year or so of conversion. In warm to hot climates fruit composition is similar to normal hand-pruned vines. Trellis costs are minimal.

Disadvantages
In cool, humid climates MPCT can lead to ripening delays associated with higher yields. While the majority of shoots may be devigorated, a few typically grow very vigorously along the top of the canopy, and so shading is likely. Experiments and commercial experience in New Zealand indicate that associated with higher yields due to MPCT there is a substantial ripening delay, increased bunch rot and reduced wine quality. The fruit zone is spread over the whole vine. Trimming of shoot tips and leaf plucking is not possible for MPCT vines. While it is easy to convert a vineyard to MPCT, conversion of a MPCT vineyard to another system is difficult and will involve crop losses, normally of one year's crop.

Management tips
Excessive crop levels can be controlled either by summer skirting or by use of mechanical harvesters to thin crops after fruit set.

Material requirements
MPCT is the simplest trellis to erect, with a single wire at 1.5-1.8 m (5-6 ft). During vine training, canes are rolled onto a lower wire to provide vine support.

Retrofitting
While establishing a new high cordon it may be desirable to retain some fruiting wood at the old head height. This can avoid crop loss during conversion.

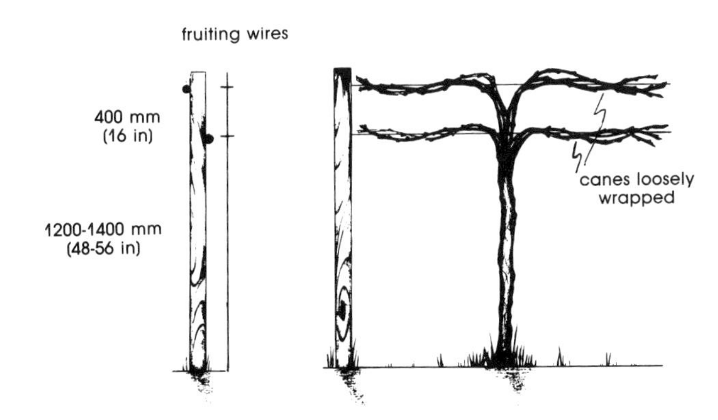

End and side view of trellis system for minimal pruning. Bottom fruiting wire is optional.

Peter Wood beside minimally pruned vines, Gisborne, New Zealand. (Photo R.S.)

Canopies of minimally pruned vines at Padthaway, South Australia. (Photo R.S.)

Lindemans Wines
Coonawarra, South Australia

Experiences with Minimal Pruning at Coonawarra

Lindemans Wines operate more than 1 000 ha (2 500 ac) of vineyard in the cool region of Coonawarra/Padthaway, the hot inland area of Sunraysia and in the Hunter River. Cultural practices are highly mechanized, particularly harvesting and winter pruning.

Substantial areas in these vineyards are minimally pruned, receiving in winter either no attention or a light trimming on the underside of the canopy. Vines are trained on a high (2 m, 6.6 ft) two wire trellis with effectively the old wood forming a framework on which the canopy rests. Grapes growing on the outside of this canopy are well exposed, resulting in good colour, loose bunches and small berries with a high skin-to-juice ratio, an important quality factor. Yield response to this system has been positive, a comparison of long term crop averages show yield increases in the range of 20 to 50%, and a much reduced incidence of biennial cropping.

In the cooler districts, fruit removal by summer trimming is necessary to regulate yield and to enhance maturity. Canopies are larger with minimal pruning requiring careful management of nutrients and soil moisture. All varieties do not respond equally to this treatment. Strong upright growers such as Cabernet Sauvignon and Sauvignon Blanc are easily managed, while pendulous varieties such as Riesling are more difficult. Vigorous vines are generally high yielding but costly to hand prune. Hand pruning can take over 100 hrs/ha (40 hrs/ac) and in contrast minimal pruning can be 5 hrs/ha (2 hrs/ac) or less. This is significant in the competitive world of wine production.

The success of this pruning method is measured by the quality of the end product. Lindemans premium and prize winning red wines from Coonawarra and white wines from Padthaway are mostly made from grapes grown on minimally pruned vines.

C H Kidd
National Vineyard Manager

Colin Kidd beside a minimally pruned vine.

Villa Maria Estate L
Auckland, New Zealan

Minimal Pruning Experiences at Auckland

This system has been trialled in Auckland vineyards since 198 The area has significant summer rainfall and humidity, wit vines showing excessive vigour. The varieties trialled wer Chardonnay, Traminer, Cabernet Sauvignon and Palomino. Th standard trained vines (VSP) were pruned to about 60 buds an leaf removal practised for the last two years.

Minimal pruning increased yields over two or three yea measured by an average of 127% for Chardonnay, 134% fo Traminer and 132% for Cabernet Sauvignon. Maturity as mea sured by sugar was however decreased, and bunch rot increa sed. We have produced wine of inferior quality from minimall pruned vines when compared to the standard trellis on the sam site. Excessive Botrytis and wet rots, uneven and delaye ripening, high acids and green 'unripe' fruit flavours are faults o the system in this wet, humid climate.

The respective decreases in sugar between standard an minimally pruned vines were for Chardonnay 1.2°, for Tramine 4.2° and for Cabernet Sauvignon 2.0°Brix. Respective averag percent bunch rot data for standard and minimal pruned vine were for Chardonnay 6% and 25%, for Traminer 4% and 20% and for Cabernet Sauvignon 0% and 10%.

Each year we assess our wine lots as a first step in blending. W found that wine from standard trained vines was graded mostl 'above average' or 'reserve', while that from minimally pruned vine was 'below average' or 'bulk wine standard'. In 1987 however, th minimally pruned Cabernet Sauvignon was graded 'above average while standard trellised wines were graded 'reserve'.

Viticulturally, vigour is not completely controlled by minima pruning and bird control is more difficult. Minimal pruning doe not seem to be a useful technique in the wet and humid Aucklan climate for the premium wine types we are producing, althoug there are benefits for bulk wine production.

Steve Smith
Viticulturist

A minimally pruned vineyard, Gisborne. (Photo R.S.)

The Ruakura Twin Two Tier (RT2T)

Background

This training system was developed especially for high potential sites such as are commonly found in New Zealand. The system was designed in 1983 by Richard Smart and first trial plantings were with Cabernet Franc at Rukuhia, near Hamilton. The system incorporates state-of-the-art knowledge of canopy microclimate and vine physiology and aims to facilitate mechanization. Since its inception, the sytem has developed several variations including reversing shoot growth direction on the bottom tier, and of planting pattern to facilitate mechanical harvest. The first commercial evaluations are now taking place in New Zealand, Australia and California.

Description

The RT2T is a horizontally and vertically divided training system. The training method relies on long cordons and spur pruning, though it could be possible to modify the design for cane pruning. **Each vine is trained to only one height**, either up or down. For the initial design, shoots in both tiers were trained upwards (RT2TA). Recent modifications have been to invert the shoot growth for the bottom tier, to bring the two fruiting zones closer together (RT2TB). This should reduce yield and fruit composition differences as found in the original design. The RT2TC has shoots from both tiers growing downwards. Shoots are trimmed short to 10-12 nodes. Shoots trained upwards are held in place by two pairs of foliage wires, those trained downwards by a single wire.

A feature of the RT2T is the retention of large node numbers per vine at pruning, up to 160. This has effectively reduced shoot vigour.

Benefits

The system has increased yields and improved quality of fruit on a high potential site. A feature is the large exposed canopy surface area, over 20,000 m^2/ha (87,000 ft^2/ac), which develops at twice the rate of adjacent VSP vines in spring. This is about the maximum amount possible to avoid excessive cross-row shading. Tiers are sufficiently far apart (1.5 m, 5 ft) so that even the base of the bottom tier receives sufficient sunlight. This large surface area gives a high yield potential, and experimental values of over 30 t/ha (12 t/ac) are recorded for Cabernet Franc in the cool climate of Rukuhia. Higher yields are possible in sunny climates. Pruning to 14 nodes/m (4.2 nodes/ft) on two node spurs gives an ideal shoot spacing and an open canopy with good leaf and fruit exposure. The open canopy gives high bud fruitfulness and discourages *Botrytis* bunch rot development. Experiments have shown that wine quality is improved over adjacent VSP vines with half the yield. We have found that leaf and fruit exposure is uniform in the canopy, which we believe is important for wine quality.

Training the arms horizontally facilitates summer trimming and winter pre-pruning by machine. Hand harvest is of course made efficient by high yields, good fruit exposure and convenient working height. At present there is no commercially available machine harvester for the RT2T in its initial configuration, though manufacturers say machine harvesting of the system is possible by modifying existing machines. If the vines are planted in the line of posts and with narrow rows (say 1.8 m, 6 ft), then the system could be machine harvested with existing machines.

The RT2T system provides for large node numbers per vine and shoots with high fruitfulness; both of these factors contribute to shoot devigoration. High early yields are a feature of the RT2T, and the extra expenses of the trellis should be recovered in the early crop years.

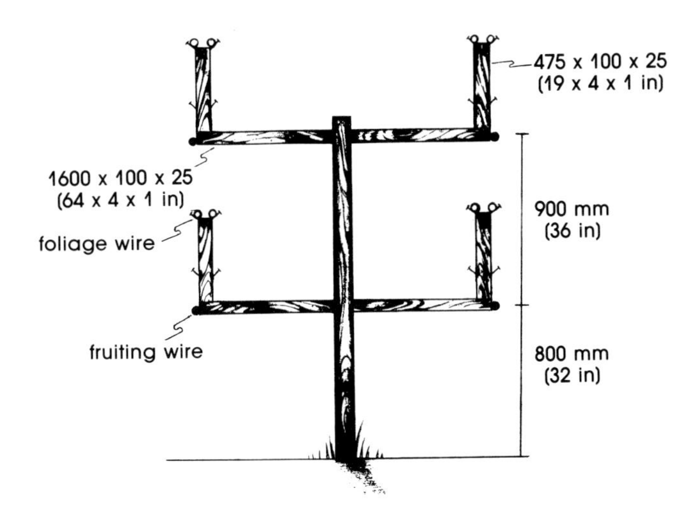

End view of RT2TA trellis constructed with central post, cross arms and vertical foliage supports.

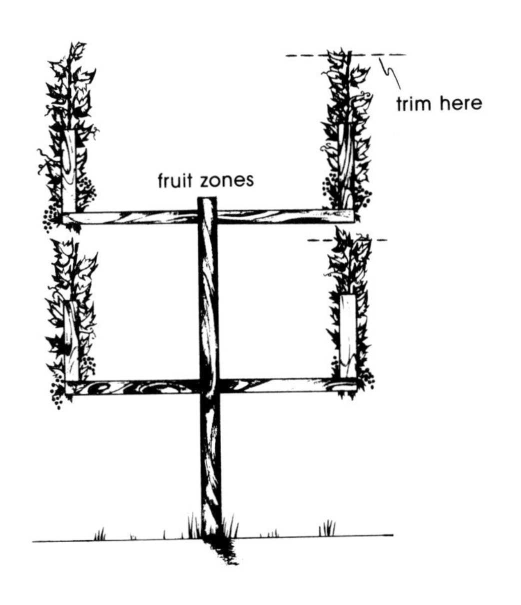

End view of RT2TA during summer showing location of fruit zone and trimming planes.

The RT2TA trellis as it appears soon after bud break. Rukuhia, New Zealand. (Photo R.S.)

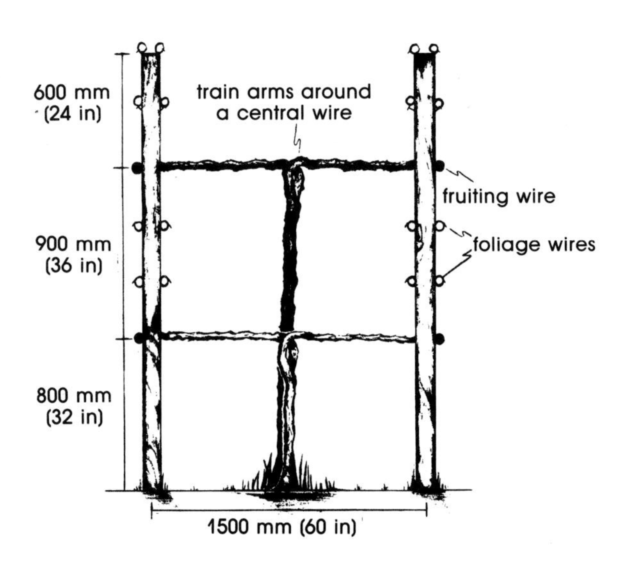

The RT2TA trellis constructed with two posts. Note one vine is trained only to one height.

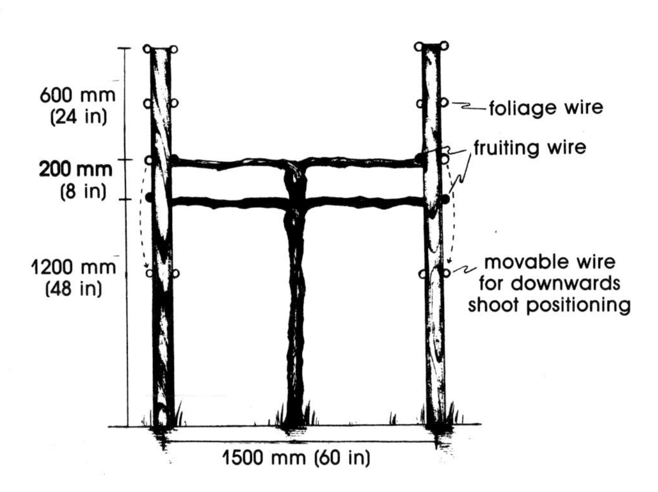

The RT2TB trellis constructed with two posts.

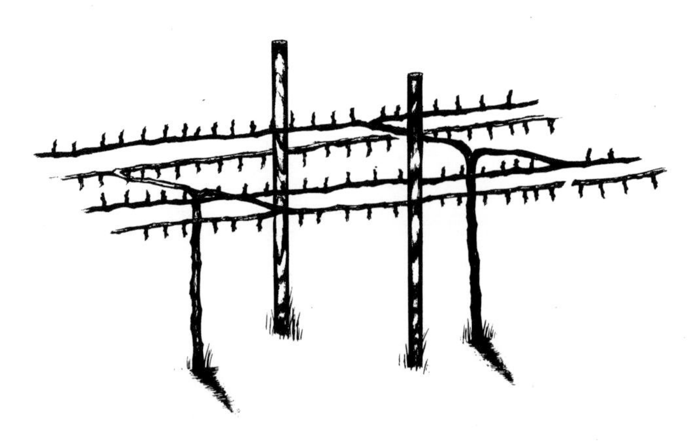

Vines trained to the RT2TB trellis. Note that the top vine is pruned to two node spurs, while the bottom vine can be pruned to downward-pointing, two to four node spurs. Top cordon is on inside of post to assist foliage separation.

Disadvantages

The RT2T system has a high material cost in posts, crosspieces and wire, and vine training costs are higher in the first two years. The RT2TA version caused ripening delays of the bottom tier, and the yield was half that of the top tier. The effect was due to shading of the bottom tier. The RT2TB form should help reduce this problem. The early growth response of lower vines converted from RT2TA to RT2TB at Rukuhia has been disappointing. Perhaps the vines were stressed in the year of conversion but caution should be exercised in following this practice. For some forms of the RT2T mechanical harvest is not presently available.

Management tips

Rapid vine growth should be encouraged in the first two years so that cordons are complete by the middle of the second growing season. We have found it possible to form 8–12 m (26–40 ft) of new cordons per vine by the end of two growing seasons for Chardonnay vines on a high potential site. This means that there will be a high potential for the first crop in the third season—which will help cover the extra cost of the system. If in the first crop there are signs that shoot growth is slowed then some cluster thinning may be necessary to avoid 'overcropping'. Only two trimmings should be required if vines are sufficiently devigorated. The first should be at about fruit set and the second about 4 weeks later when fruit is pea size. Do not let shoots from one tier grow into another. For the RT2TB version, position shoots downwards in two operations as described for the Scott Henry. Clusters can be thinned after set to reduce crop on the bottom tier and to reduce ripening differences between the two tiers.

Mechanical pre-pruning is easy as the canes are not thick, are widely spaced and all vertical. A machine wth two cutter bars per tier is efficient—one cutting at 2–3 node height, the other one cutting canes off the foliage wires. Hand follow-up pruning to obtain desired bud level is possible at almost walking pace as the spur density is low. The total pruning operation for all four tiers should take less than 30 hrs/ha (12 hrs/ac).

Material requirements

Post height of 1.8–2 m (6–6.5 ft) out of the ground is required, and the posts should be 750 mm (36 in) into the ground. Using the centre post and crosspiece construction, the crosspieces for fruiting arms should be 75 × 50 mm (3 × 2 in) timber or equivalent, and crosspieces for foliage wires 75 × 25 mm (3 × 1 in). Row spacing should be 3.6 m (12 ft) minimum. Alternatively, the RT2T can be constructed using two lines of posts at narrow spacing (say 1.8 m, 6 ft) and then crosspieces are not required. If vines are planted in line with the posts, the RT2TA then is effectively a narrow-row version of TK2T, and the RT2TB equivalent to a narrow-row version of Scott Henry. Vine spacing within the row needs to be wider to allow for high node numbers per vine, at say 3–4 m (10–13 ft). For each tier with shoots growing up, there is required one fruiting wire and two pairs of moveable foliage wires. For downward growing shoots you will need for each tier one fruiting wire and 1 or 2 foliage wires. The RT2TA requires therefore four fruiting wires and 16 foliage wires per row, while the RT2TB requires four fruiting wires and 10–12 foliage wires.

Retrofitting

The RT2T can be retrofitted to rows of 3.6 m (12 ft) or more width, and where the head height is about 0.8–1.0 m (32–40 in). Remember to avoid summer trimming the year before conversion, as you will need long canes to form new cordons. For example, with vines planted 2 m (6 ft) apart, you will need four canes at least 2.8 m (9 ft) long to fill the cordon wire.

Department of Viticulture and Enology,
University of California
Davis, California, USA

Experiences with RT2T in California

Three 25-vine rows of Sauvignon Blanc vines grafted to AXR#1 rootstock were planted in the spring of 1985 at the University of California Experimental Vineyard located at Davis, in a very deep Yolo clay loam soil at row and vine spacing of 3.6 × 2.4 m (12 × 8 ft) with row direction running east to west. The three rows were trained to an RT2TA trellis. One row of 25 vines trained to a five wire, 2 m (6 ft) high vertical trellis (VSP) was also planted at the same time using the same within row vine spacing as the RT2T vines and served as a control for comparison. The vertical and RT2TA trellis systems had 2.4 and 9.6 m (8 and 32 ft) of canopy length per vine, respectively.

RT2T vines averaged 136 shoots/vine with a shoot density of 14 shoots/m (4.2 shoots/ft) compared to 58 shoots/vine and 24 shoots/m (7.3 shoots/ft) for vertically trellised vines. Average shoot length and leaf area per shoot of vertically trellised vines was 1.45 m and 5037 cm^2 respectively, compared to 0.6 m and 2140 cm^2 respectively for RT2T vines. The average internode length of RT2T shoots was significantly less than for shoots of VSP vines.

The first crop from the RT2T in 1988 was 32 t/ha (13 t/ac) compared to 11 t/ha (4 t/ac) for VSP vines. Yields for the second year were 45 t/ha (18 t/ac), and 18 t/ha (7 t/ac) respectively. The number of clusters per vine for the vertical and RT2T trellis systems ranged from 65 to 77 and 194 to 242 respectively. Cluster weight and berry weight were generally sightly smaller on RT2T vines than vertical vines.

RT2TA fruit has taken one to three weeks longer to reach 22°Brix than fruit from vertically trellised vines, no doubt due to the heavier crop. The crop load for RT2T was probably excessive relative to leaf area per shoot. The pH, titratable acidity, potassium and malate levels of RT2T fruit at harvest have consistently been lower than fruit from the vertical trellis. Wine made from RT2T fruit was lower in pH, titratable acidity, malic acid and potassium, and less vegetative in character than wine made from vertically trellised vines. The RT2T wine was a little more golden in colour than wine made from the vertical trellis. On a 20 point score, an expert taste panel rated the RT2T wine 13.9 compared to 13.1 for the vertical trellis wine. Wine quality was certainly not reduced despite yields being more than double.

Growth, crop yield and fruit composition of RT2TA vines trained to top cordons differed greatly from RT2TA vines trained to bottom cordons. Yield of RT2T trained high averaged 40 to 100% higher than RT2T trained low. Fruits from the high cordon were also lower in pH, potassium and malate compared to the low cordon fruits. These differences can largely be accounted for by considerably higher sunlight exposure of leaves and fruit of the upper cordon compared to the lower cordon with east-west row orientation. Preliminary data from a new planting of vines on RT2T trellis with row direction running north-south revealed much smaller differences in light exposure between the upper and lower canopies.

Mark Kliewer

W. Mark Kliewer
Professor of Viticulture

The RT2TA canopy with vigorous Sauvignon Blanc at University of California, Davis. (Photo R.S.)

The original Ruakura Twin Two Tier vines. The canopy of vigorous Cabernet Franc vines trained to RT2TA produces over 30 t/ha (13 t/ac) yet with excellent leaf and fruit exposure. Rukuhia, New Zealand. (Photo R.S.)

Mark Kliewer

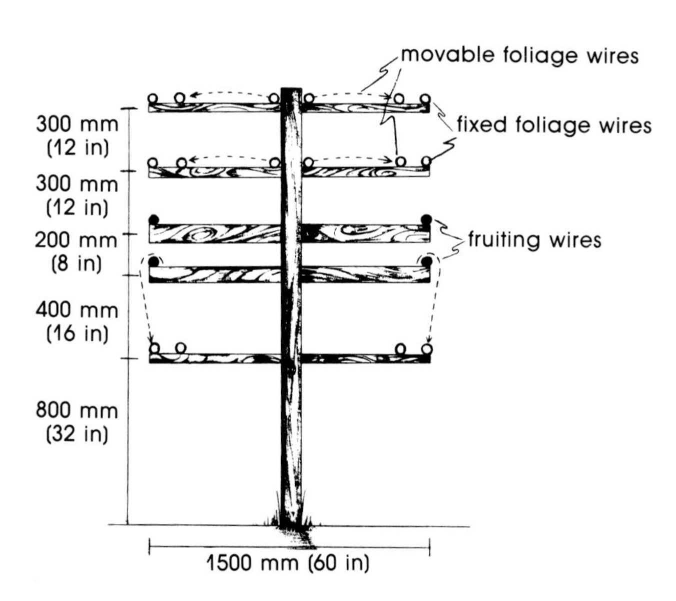

End view of trellis for RT2TB using a central post and cross arms.

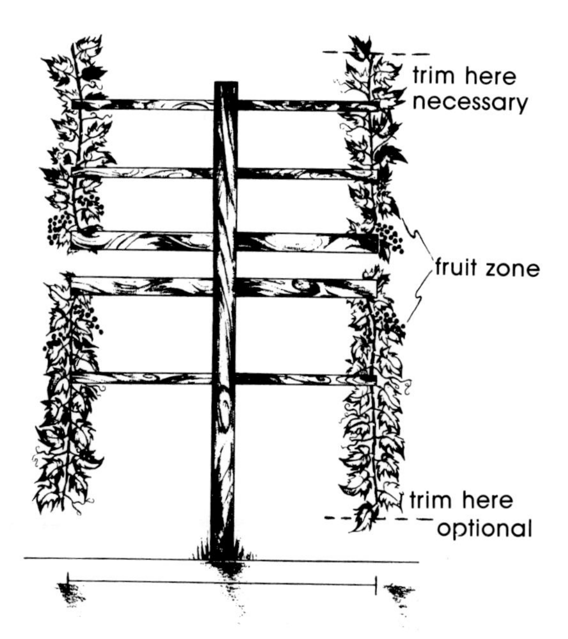

The appearance of RT2TB vines in summer showing fruit location and trim planes.

MAF Ruakura Agricultural Research Centre
Hamilton, New Zealand

Experiences with RT2T at Ruakura

Four years of harvest data have now been collected from the Cabernet Franc trellis trial at Rukuhia near Hamilton. Results have shown that yield from the RT2TA is consistently twice or more that of the New Zealand standard (VSP). Features of the RT2TA that lead to an increase in yield include a larger canopy surface area, higher percent budburst, more clusters per shoot and berries per bunch. Not only is yield improved but so also is wine quality. Although at harvest RT2TA tends to have slightly less sugar (up to 0.9°Brix) than the standard, it also has lower pH and lower incidence of bunch rot and these latter differences carry over into the wine. In 1988 and 1989 industry judges clearly preferred RT2TA wines to those made from VSP fruit, judging that RT2TA wines had better colour, more fruit character and were fuller on the palate.

Management of the RT2TA requires shoot positioning to be carried out during spring and early summer in order to maintain the shape of the canopy. Summer trimming is required to keep the two vertical curtains discrete, but the RT2TA is trimmed no more often than is the VSP. The open canopy ensures good spray penetration and pickers favour the RT2T as the fruit is easily accessible and both tiers are at a convenient height.

Big vs small

Design of the trial has meant that the RT2TA has two vine sizes, 'big' vines (12 m or 39 ft cordon, pruned to 160 nodes) and 'small' vines (6 m or 20 ft cordon pruned to 80 nodes). Big vines show devigoration, resulting in reduced cane interode length and weight, and fewer and smaller laterals. Overall, the big vines have a better balance between vegetative and fruit growth than do either the VSP or the small vines on RT2TA as is shown by higher yields per shoot, higher yield to pruning ratio and lower cane weight.

Towards RT2TB

Cross-row shading has led to yield and fruit composition differences between the top and bottom tiers of RT2TA. The top tier is characterized by having twice the yield of the bottom and higher sugars, although pH and TA values are essentially the same. These differences have led to the decision to lift the bottom cordon to bring the fruiting zones within 25 cm (10 in) of each other (RT2TB). Conversion was initiated in winter 1988 and new cordons were formed while crop was maintained on the old cordon. The new cordon lengths required (12 or 6 m, 3.9 or 20 ft) were easily developed within one growing season. During winter 1989 the old cordon was removed and the new one shifted to its final position. A 2% solution of hydrogen-cyanamide was applied 2.5 weeks prior to the budburst to ensure even bud break. Bud break was in the order of 90%. Subsequent management was similar to that of the spur-pruned Scott Henry system with the selection of downward pointing spurs at pruning. During the first crop year (1990) on the new cordon, yields were still less than the top tier, though as the bottom cordon matures and spurs develop we expect this difference to decrease. Differences in maturity between top and bottom tiers were reduced at harvest due to conversion to RT2TB, with the top tier being 1.3°Brix higher and with 0.7 g/L less titratable acidity. In 1991 the yield and vegetative growth of the bottom tier was reduced, and perhaps there has been a negative effect of conversion. Further studies are required.

Joy Dick
Research Technician
(Joy Dick's photograph appears on page 7)

The Economics of Canopy Management Practices

In view of the projected international audience of this handbook, we have refrained from giving detailed costs of canopy management procedures in local (New Zealand) currency. Furthermore, economic data is unfortunately soon modified by inflation, and the cost of components and labour and installation can vary widely from country to country, even region to region. Therefore, we do not present budgets of suggested procedures here, but rather in general terms discuss the principles we have developed from vineyard costings for clients in New Zealand, Australia and the USA. The information required for readers to do their own budgets is presented elsewhere in this handbook, including post lengths and spacings, wire required and labour requirements, etc.

Canopy management techniques vary widely as to the cost of introduction and their economic benefits. At one end of the spectrum is a simple procedure such as leaf removal in the fruit zone, which, especially if done by machine, is a low cost operation. Similarly, the projected benefits are not huge, unless, of course, you save a crop because Botrytis bunch rot is avoided! The extra costs of leaf removal are incurred in mid-summer, and the economic return can be gained a month or so later at harvest. At the other end of the spectrum there is the more substantial investment in complex trellis systems such as the RT2T for a new vineyard. The greater capital investment will not give increased economic benefit for several years, typically after several crops. In between these two extremes there is the common proposal to retrofit a trellis system. Here, the investment is moderate and the economic advantage is normally gained within one or two crop years.

Partial budgets

Partial budgets can be done with a hand calculator. One simply compares the costs and returns from a proposed management change with the existing procedure. For example, to investigate the economic benefit of leaf removal, list the extra costs which are: machine costs (capital value amortized over the life of the machine, and annual running costs), tractor and operator costs, or alternatively, labour costs if leaf removal is done by hand. Similarly, the projected extra income will include an assessment of any yield increase (which could be substantial in Botrytis-susceptible vineyards), the projected fruit value and any quality bonus which may occur.

For example, a New Zealand analysis in 1989 for Sauvignon Blanc vineyards indicated a net benefit of almost NZ $4000/ha ($1600/ac) for adoption of machine leaf removal costing $140/ha ($56/ac) where 5 t/ha (2 t/ac) was saved from Botrytis attack. Such a benefit is based on field experience. Where the projected yield increase was smaller, 1.2 t/ha (0.5 t/ac), the net benefit was $1330/ha ($532/ac). Similarly, a partial budget for trellis retrofitting would involve the additional costs of components (posts, post extensions, cross arms, wire, etc.), installation and shoot positioning. The additional income could be calculated from the projected yield increase, value of the fruit and any possible bonus for quality improvement.

Development budgets

These budgets require more complex calculations and are typically done using computer spreadsheets which offer the advantage that different options can be evaluated simply. Development budgets cover the first 10 to 15 years of the vineyard's life, and include all capital and annual operating costs. An important component of such budgets is the debt servicing of borrowed capital. High interest rates can make some options very expensive. When run on spreadsheets the effect of variation in yield and price can be easily accounted for, which allows for a sensitivity analysis of the budget. As an example of a development budget, a 1989 comparison of the establishment of a VSP and RT2T vineyard was made in New Zealand using actual yield data. The RT2T cost almost $7800/ha ($3120/ac) more to establish due to more posts, wire and training costs. The yield advantage of the RT2T (13 t/ha, 5.2 t/ac) led to an annual extra income of $9700/ha ($3870/ac) after maintenance costs were allowed for. Without debt servicing, the break-even point was year 6 for both systems, and with interest at 18% it was year 9. However, in both cases the RT2T system was the better investment, with the extra yield much more than compensating for the extra capital cost.

Economic responses to canopy management

A few general principles emerge from numerous financial analyses carried out for vineyards in several countries which have adopted canopy management practices. These are:

- High yielding vineyards are generally more profitable than low yielding ones, although of course grape price is also an important determinant of gross income. This situation may be altered if wine rather than grapes is sold;
- Vineyards with a high degree of mechanization are generally more profitable. In particular, the labour costs of pruning and harvesting can become major annual expenses;
- The extra cost of a more elaborate trellis is repaid when the vineyard vigour warrants its use. This is because the 'excess' vigour can be converted to increased yield;
- Vine planting material can be a major establishment cost, especially for grafted vines. This can mean high density plantings are much more expensive;
- Retrofitting a vigorous vineyard on a simple trellis system to a divided canopy typically gives a one or two year payback.

Kate Gibbs doing a vineyard budget for a client on a portable, lap top computer. (Photo B.W.)

Common Questions about Canopy Management with Answers

Fruit showing sunburn. Note yellow and shrivelled berries. Barossa Valley, Australia. (Photo R.S.)

Bud position controlling early shoot growth direction for a horizontal spur. Cabernet Franc, Rukuhia, New Zealand. (Photo R.S.)

A poorly shoot positioned vineyard. It is difficult to determine the trellis system used. Somewhere in California. (Photo R.S.)

Listed here are questions we are most commonly asked.

Won't fruit get sunburnt if it is well exposed?

Generally fruit does not get sunburnt as long as the clusters have been well exposed for a long time, preferably since flowering. They become 'hardened' much like humans are able to build up a suntan to avoid sunburn. Hot, sunny, calm weather leads to sunburn, so avoid exposing previously shaded fruit under these conditions. This is one reason why we recommend that foliage management begins early in the season.

If I remove vines as a means of devigoration, won't vine vigour increase as roots fill the vacant space?

It is difficult to categorically answer this, but our limited research experience and evidence from commercial vineyards suggests not. Studies in Australia and New Zealand have shown that root growth can in fact be slowed with large yields per vine, and this probably causes the vines to remain devigorated.

Can I train the same vine to two different heights, as for the TK2T, Scott Henry or RT2T trellis?

No! Eventually the top cordon will dominate the lower ones, by a combination of shading and gravimorphic effects.

Why is the timing of shoot positioning so critical?

Early in the season, shoots can be easily broken out of their sockets. Strong winds will often do this. They become more securely attached just before flowering. At about flowering, tendrils become 'thigmotropic' (touch sensitive) and bind the canopy together, making it difficult to position shoots. So there is a 'window' of about three weeks before flowering when shoot positioning should be done: too early, and shoots break; too late, and tendrils make it difficult.

Don't shoots want to grow upwards after you turn them down?

Yes they do, and the tips will gradually grow back up through the canopy if allowed. Shoots should be trimmed soon after turning down. Shoots are devigorated by growing downwards and require less trimming anyway.

Why do you suggest spur pruning over cane pruning?

There are several reasons. Firstly, it is easier to arrange shoot spacing on cordons than on canes, since there is typically poor bud break in the middle of canes. Shoot growth is also more uniform. Secondly, spur pruning can be easily mechanized and cane pruning cannot. A third reason is that you cannot produce a 'big vine' effect with cane pruning, because you need long cordons. However the disease phomopsis requires more careful control.

How quickly should cordons be established on young vines?

As quickly as possible on high potential sites. By encouraging rapid growth in the first two years (with adequate water and nitrogen and good weed control), it should be possible to grow about 10 m (33 ft) of cordon per vine in two years. This requires constant training, especially in the second year. Use hydrogen cyanamide to ensure good bud break, and high yield in the first crop year will follow.

Won't this practice of early cropping cause 'overcropping' and affect the vine's health?

Overcropping can be recognized when shoots stop growing early in the season before flowering, and where there are plenty of bunches. If this occurs, remove bunches early to bring the vine into balance. Our experience suggests, however, that overcropping is unlikely provided there was adequate healthy foliage in the previous year. More common is the practice of restricting cropping in the first few years by fruit removal. It is likely that this practice makes future vigour problems worse by stimulating root growth.

I am planning to topwork my vineyard to change varieties. Can I combine this with trellis conversion?

Yes, this is an excellent idea. Normally the year following grafting produces little crop and very strong shoot growth. So new trellis systems can be quickly implemented and cordons filled.

How much crop loss can I expect with vineyard conversion to an improved trellis?

Normally you should not have **any** crop loss, but in fact a crop increase. For vigorous vineyards, as long as you leave more nodes at winter pruning during conversion than you previously had, then yields should increase. This is why it is so important not to trim hard the previous summer, as you typically want to use long canes during conversion to make sure you get the right node number. However, there are some instances where you need to cut cordons or canes off in retraining, and this may cause a yield loss for one year.

How can I arrange to have spur positions correctly spaced when establishing a new cordon?

To achieve about 15 shoots/m you need about 7.5 spurs/m, or one spur each 130 mm (5 in). Select medium vigour canes for your new cordon, and the internode length will be about 70 mm (2.8 in). Then, the following winter by thinning out to every second shoot (assuming 100% budburst), you have about the correct spur spacing.

How critical is the timing of trimming?

Trimming is easiest to carry out when the shoots are held firmly in place by foliage vines, rather than being long and falling in different directions. Trimming early and often makes the job much easier. Normally trim for the first time soon after flowering, and the second when berries are small pea size. Do not allow shoots to grow longer than 400 mm (16 in) beyond foliage wires as they will fall over and be difficult to catch with the trimmer.

Should I plan to use the same trellis system throughout my vineyard?

Use of a single system makes the vineyard easier to manage but is likely to make fruit composition similar. Wherever it can be accommodated we suggest using different systems in the vineyard to create differences in fruit composition and ultimately more complexity in the wine. Do not forget that different trellis systems can interact with seasonal conditions, so that one trellis system is unlikely to be the best for all seasons.

I want to use Scott Henry but my posts are too short. What can I do?

You may not have to repost your vineyard. You can attach a post extension which will be sufficient to hold the foliage wires. These can be made of wood or steel.

I want to convert my vineyard to GDC with new posts, but am afraid that my end posts will be too weak. The posts are only 150 mm (6 in) wide and 600 mm (2 ft) in the ground. Should I change them?

You will definitely need to strengthen your end assembly. The existing posts probably cannot handle tighter wires, let alone increased yield. If the headlands are wide enough, the simplest solution is often to add tie-backs.

When I improve my trellis system will I need to alter fertilizer and irrigation programs?

Yes, there will be increased needs for both nutrients and water as vine growth and yield increase. You should check for altered fertilizer needs by tissue analysis. Irrigation needs will increase, typically in proportion to increased canopy surface area.

My neighbours say my vines will die if I use a 'new fangled' trellis system. Is this true?

Traditional and conservative vine growers have a lot of difficulty in accepting new ideas. You will often find that they are being defensive and do not want to accept that their ideas may be wrong. There are only a few things that kill grapevines. New trellis systems are definitely not amongst them!

Shoots on a very fruitful cane at flowering. This high degree of cropping will probably require some thinning. South Africa. (Photo R.S.)

David Jordan appreciating well spaced spurs and hence shoots, RT2T. Rukuhia, New Zealand. (Photo R.S.)

A galvanized steel post extender manufactured by Vocol Engineering which allows conversion to VSP or Scott Henry of a low post height vineyard. Hastings Valley, Australia. (Photo P.M.)

Constructing Trellis Systems

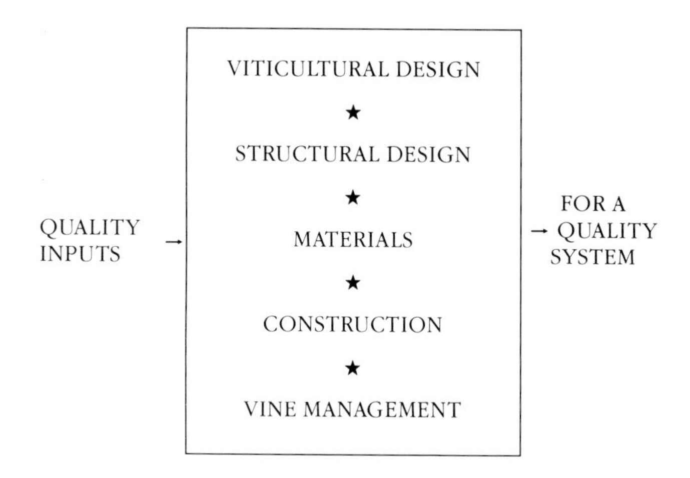

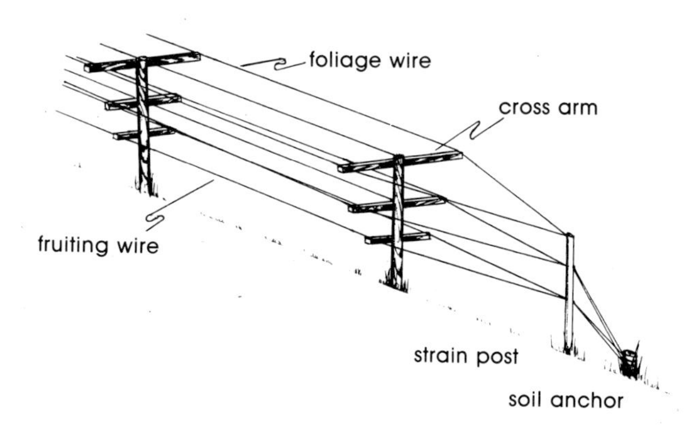

The U or lyre trellis to show component parts of a typical trellis.

Mike Robinson (in shorts) demonstrating loads on a vineyard trellis with concrete blocks. Sonoma Workshop, July 1989. (Photo R.S.)

A Trellis for Your Vineyard

The preceding notes have focussed on the viticultural aspects of trellis design. They have been prepared to help you choose the type of trellis to suit your vineyard. We now turn to the question of how to achieve that type with a strong and durable trellis **structure**—and at a price to suit your pocket. Remember that trellis costs occur at many levels. In addition to the obvious material purchase and construction costs, consider the time spent on vine training, pruning and harvesting. The trellis structure can affect these costs. The cost of repairs is also a consideration—particularly with time and production lost during those repairs. The notes that follow have been organized to show the key factors which should be considered when choosing and building a trellis structure.

Trellis components

To discuss trellis structures, we must first define the component parts of a typical trellis. The drawing shows a U or Lyre trellis constructed using a central post. Other methods of construction are possible, but this one has been chosen to illustrate the key components and dimensions. Note the key features: a soil anchor and strain post comprising the end assembly, an intermediate post, cross arms, fruiting and foliage wires.

Loads on the trellis

Crop load: Crop load is the total weight of fruit, shoots and trunks or cordons that are supported by the structure. The extent of crop load not only depends on the yield, but also on the shape of the support structure, and the methods of training, pruning and managing the crop. A trellis for a vineyard is required to support virtually the total crop load (whereas a tree crop partially self-supports its load). In your vineyard this will be up to 10 kg per metre (7 lb/ft) of each fruiting wire.

The load distribution on a trellis is seldom uniform. On the individual vine uneven load distribution may come, for example, from the concentration of fruit at the end of the cane. On the trellis posts, shading of one side of the vine, vineyard edge effects, and harvesting when one side of the row is picked first can all result in unbalanced loading and lead to overturning problems on trellis posts.

Other loads: Trellises should be able to support a number of secondary loads including:

- Accidental impact during vineyard operations. Clearly it is not possible to totally 'idiot-proof' structures, but they should be able to withstand normal wear and tear.
- Wind, particularly for exposed vineyards.
- Snow or ice loading. Fortunately these normally occur out of the growing season.
- Mechanical harvesters. These obviously impart a lot of vibration to the trellis, and this is bound to identify any weak spots at the worst possible time!

Load Transfer Through the Trellis

This section introduces the general concepts which describe the structural workings of the trellis. These concepts are discussed in more detail later as the selection of individual components is considered.

The load is transferred through the trellis via the following steps. Note that if the parts of any one step are too weak, the whole system fails. Because it is the main component of load, only fruit load is shown, but the same steps apply for any load.

Step 1. Fruit load
The small arrows on the diagram show the fruit weight hanging down from the fruiting wires. This puts tension in the wires in the same way as for a suspension bridge, and indeed the wires will sag in the same way as suspension bridge cables do. Tensions in a loaded fruit wire are typically about 1500 N (330 lbf). Tension increases with load and span, and decreases as the sag is allowed to increase for a loaded wire. A common sag is about 175 mm (7 in).

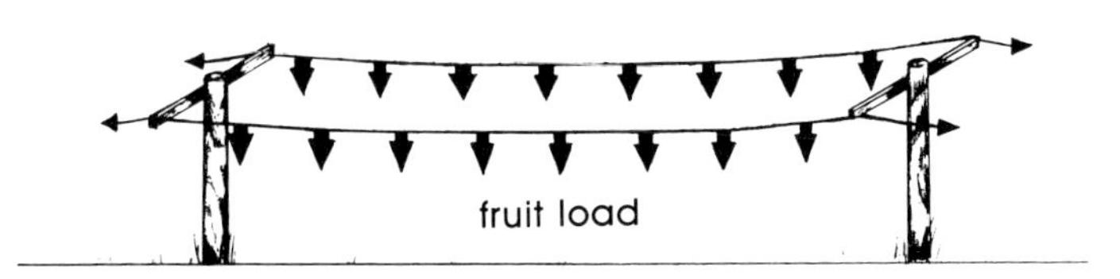

Arrows represent fruit load on the trellis wires, causing tension in the wires.

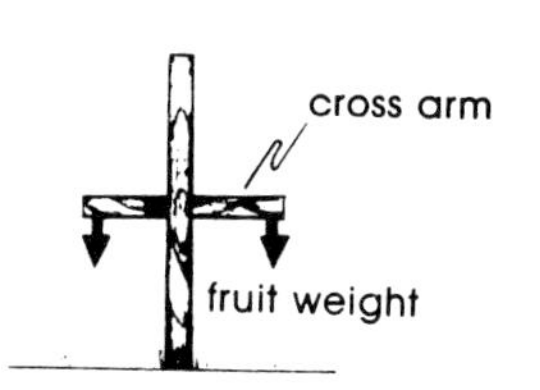

The fruit load is transferred by the wires to the cross-arms.

Step 2
The total fruit weight in each space between the intermediate posts is held by the cross-arm and its connection to the post, or the staple holding the wire to the post when there is no cross-arm.

Step 3
The downward load is finally transferred from the post into the ground. For virtually all soils and posts, the area of the base of the post and the friction of the post on the soil are sufficient to stop the post sinking.

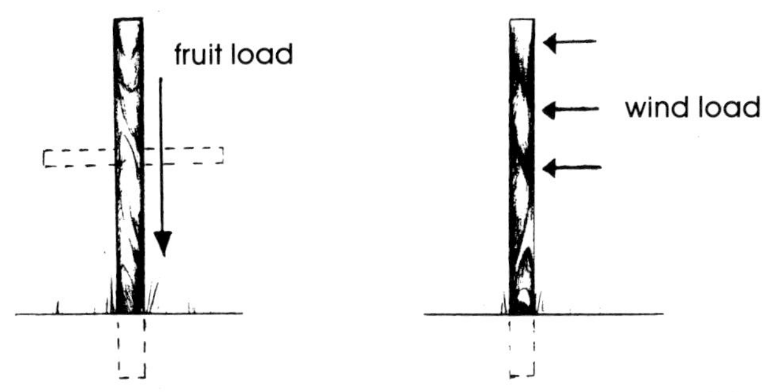

The load is transferred to the ground.

Wind load acts on the side of the post.

Windload: In practice, the most common cause of post failure is the leverage caused by loads such as wind blowing side-on onto the canopy. These loads are much smaller than those due to the fruit, but are significant because they act over the total height of the post, thus causing more leverage.

Often this effect is exacerbated by water running down the back of a loosened hole around the post causing the post to fall over rather than break. For this reason, if practicable, locate your irrigation emitters away from the post.

Step 4
Tension in the wire is held at the end of the row by the end assembly. The shapes and sizes of end assemblies differ greatly but the basic purpose never changes. The purpose of an end assembly is to provide an anchorage point for the wires. It is the link between the wires and the soil. It needs to be strong and normally needs to be stiff, i.e. it moves only a little when the wires are tensioned. The resistance to movement comes firstly from strength as a framework, in the same way that a truss in a roof must be strong. The second aspect is that of anchoring in the soil. To achieve this the framework must spread the load over a sufficiently wide contact area of soil so that the soil will not move under the load. This is the reason why larger diameter posts are used in end assemblies. The firmness and stickiness of the soil govern the contact area required.

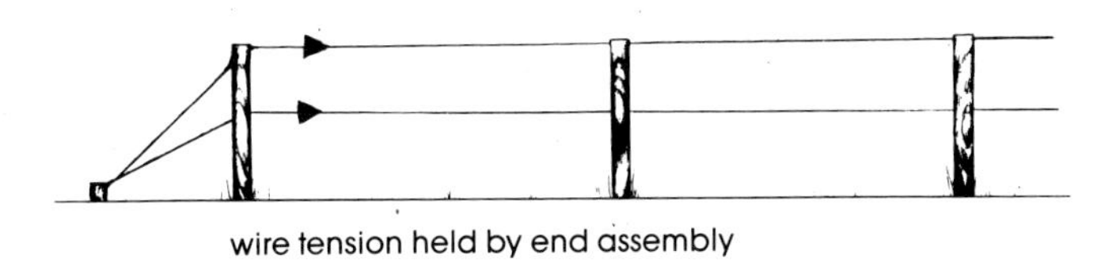

The end assembly must hold tension in the wires.

A collapsed U trellis with high vigour vines. Somewhere in Australia. (Photo R.S.)

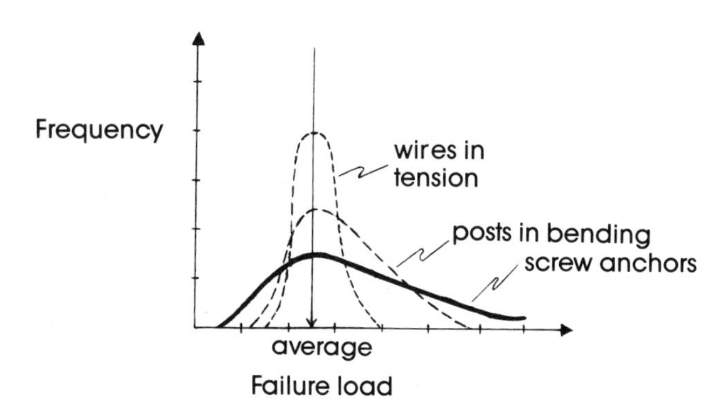

A schematic illustrating the likely spread of failure loads of trellis components, all normalized about the same average value.

A collapsed end assembly. California. (Photo R.S.)

A metal cross piece unable to withstand mechanical harvester damage. Hunter Valley, Australia. (Photo R.S.)

Key Strength Areas of the Trellis

It is always possible to build a trellis so strong that it will never break, move or fall over. However, this would be very expensive because in every batch of posts or pallet of wire or area of soil etc., there will be some weak spots. For instance, if a hundred 150 mm (6 in) plate screw anchors were installed in a vineyard and tested, five would pull out at say 1800 N (400 lbf), five at 15000 N (3300 lbf) and the rest somewhere in between. To a lesser extent, this same effect occurs with all trellis components. Obviously, to guarantee strength, the average strength should be appreciably larger than the minimum required. To do this over the whole structure is usually uneconomic. Instead we consider the consequences of failure. If something breaks . . . how will it break, what will the crop losses be and how difficult will it be to fix? The balance between cost and frequency of repair and cost to make it stronger in the first place leads to economic trellis design.

Zones of importance
To answer these questions, trellis structures can be divided into three zones of importance in terms of strength and durability.

Zone 1, very important: Places where failure could lead to appreciable crop loss and disruption of harvest in that block, and where repair is difficult and time consuming. Examples are end assemblies, and in particular, the back anchor posts.

Zone 2, important: Places where failures cause localized crop loss and repair is relatively straightforward. Examples of this are failures of intermediate posts, fruiting wires or cross arms.

Zone 3, unimportant: Places where failures are of nuisance value only, and so are easily repaired, such as with foliage wires.
Let us consider some structural components using the zone classification philosophy.

End assemblies: These should ideally be tested before being used. It is tempting to use a similar assembly to others used in your district, but, before doing so, check that both the soil type and the loads will be the same. In particular if you are using improved trellis designs you will need strengthened end assemblies which may be more substantial than those in common use. It is clearly an area where a little extra care will not cost much but produce very worthwhile returns in the longer term because of the disruption caused by end assembly failures.

Intermediate posts: These are a significant part of trellising costs. It is natural to wish to use shorter posts, weaker posts or fewer posts for reasons of economy. Note that posts that are too short may not permit construction of trellis systems with correct dimensions, so yield and quality can suffer. In addition, increasing the post spacing increases wire tensions, wire sag and end assembly loads so savings here could be short-sighted. Typically, spacings of 6 m (20 ft) are used in vineyards.

Foliage cross-arms and wires: These components are easy to repair so they are more suited to making do with the bare minimum of materials if the budget has to be trimmed.

Soil: Perhaps surprisingly, soil is a key component of any trellis structure. As was illustrated in the load transfer section, all the fruit and other loads must eventually be transferred to the soil as pressure between the post or anchor and the soil. If the pressure is too great, the soil will compress, anchors will pull out, or posts lean over. To avoid this happening to your trellis, investigate the strength of your soil so that post embedment and anchor sizes can be designed to suit. Note, though, the important soil layer is that depth at which the anchor is pushing. This is discussed in more detail later.

Trellis Durability

The trellis you build today will soon be needed for training the vine. The big loads, however, will be several years away when the vines have matured, normally in years three to six after planting. How will your trellis have fared in the intervening years? Without proper attention to rust proofing and rot prevention, there may be cause for concern.

Timber components
Some timbers are naturally durable. These are mostly the hardwoods, and particularly the heartwood of such trees. Most timber used for trellises is softwood. Species such as pine are economical, easily worked and (usually) available. To ensure an economical life though, they must be treated to prevent decay and insect attack. Most countries have treatment standards which are effectively a guarantee. When the timber is used in stated situations then it will remain in sound condition for a number of years. For example, in New Zealand 'H5' is the standard for horticultural timber in contact with the ground (e.g. posts), 'H4' for agricultural posts and 'H3' for exposed timber not in contact with the ground, such as cross-arms. The reason for the variation between agricultural and horticultural posts is that in a vineyard the soil is usually more fertile, watered more regularly and usually better draining. Together these factors lead to early rotting of inadequately treated posts.

Posts treated to the H5 standard in New Zealand will last at least 25 years. Note that for post treatment such standards in fact specify minimum preservative levels in the outer fibres of the post. This is particularly important with round posts as the preservative travels easily up the sapwood but not so in the less porous heartwood. Peeler cores which are essentially all heartwood are therefore very difficult to adequately treat. If you cut into the post you could then be cutting through the treatment 'envelope' and so causing early rot. In general any cut deeper than about 25 mm (1 in) should be spot treated with a paint-on preservative.

There are many different types of timber preservatives. The most common belong to the CCA (copper-chrome arsenate) family. This very effective chemical has one drawback. The copper reacts with zinc, which is the key element in galvanizing. Contact between the two can therefore lead to the treatment salts stripping galvanized layers off the wires. It is generally only a problem with posts that have been treated less than a month before they are used in the trellis, as after this time the most readily leached treatment salts have disappeared.

Steel components
The most common rust proofing for steel is galvanizing. Galvanizing is essentially the placement of a zinc coating over the steel. The zinc slowly oxidizes away which stops any nearby steel from oxidizing (oxidation is rust). This is called sacrificial protection, and extends up to 6 mm (0.25 in) across a bare steel surface away from the edge of the galvanizing. Apart from being relatively economical, the major advantage of galvanizing is that because of sacrifical protection, worn spots do not necessarly become rust spots.

Galvanizing is also the best method for ensuring durability of steel wires. Systems such as plastic and aluminium coating are only suitable if extreme care is taken not to damage the coating during erection and during its working life. This is because any holes in the coating quickly lead to rust. For galvanized fencing wires, to ensure the galvanizng is of good quality and will not flake off when you tie the wire, look for a manufacturer's warranty of compliance to a code.

In aggressive acid soils, galvanizing is less effective so a robust paint system that will not chip off when knocked is best. An ideal composite system for such conditions is to use galvanized stakes, but to paint the lower half that will be in the ground.

A failed tie back assembly. (Photo M.R.)

Non-treated timber posts and rails. Australia. (Photo R.S.)

Common Californian sight of a steel stake for each vine. Some newer vineyards are using wooden intermediate posts every three vines. (Photo R.S.)

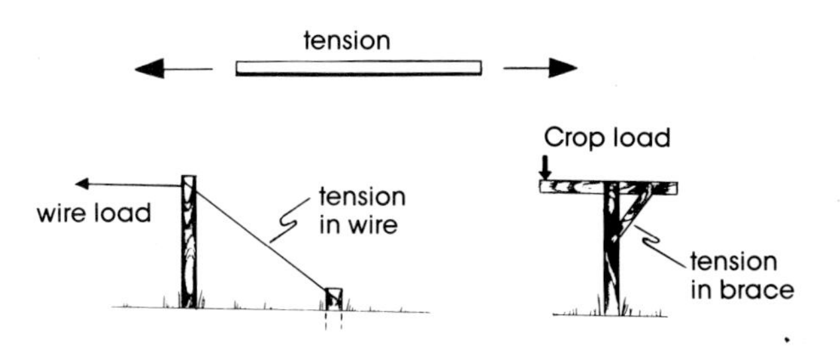

Vineyard situations where tension is experienced.

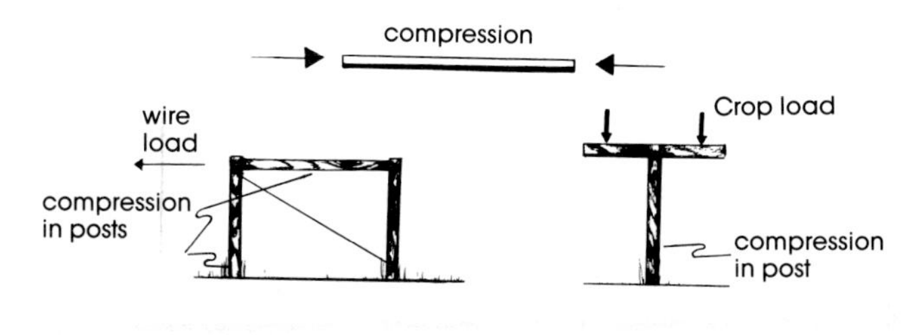

Vineyard situations where compression is experienced.

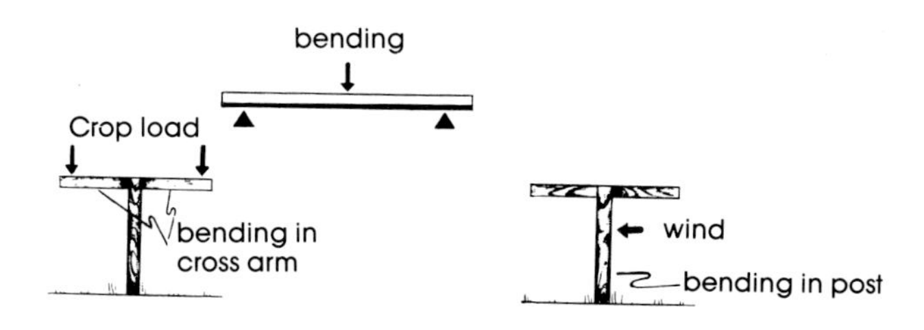

Vineyard situations where bending is experienced.

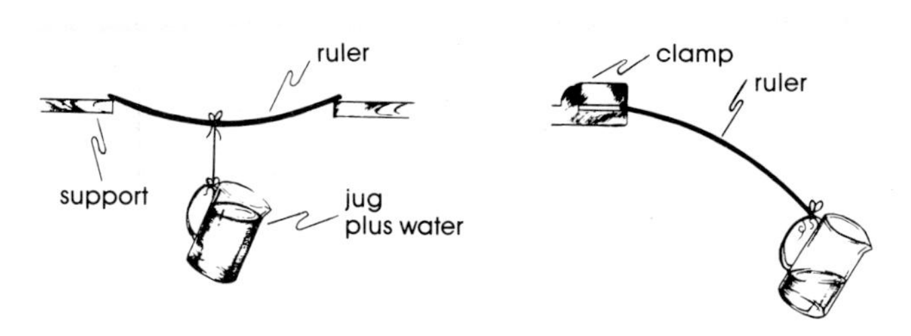

Demonstrations of bending. Notice greater bending in the ruler when supported only at one end.

Trellis Component Strength

A trellis is made up of many components, each of which has a certain strength. The strength is dependent on the size, shape, material and orientation of the component. To survive, clearly a component must be stronger than the load on it. There are three relevant load types.

Tension

When a component is stretched it is in 'tension'. Wires are the most common tension member, but others such as diagonal braces on an intermediate frame can also be in tension. A tension failure invariably occurs at the connection points—at knots in wires, or at nail/bolt points on braces.

Compression

Compression is the opposite of tension. In a trellis, the most common compression situation is a post supporting a downwards crop load. When a member is squeezed it resists the load to a point where it suddenly buckles out sideways. To observe this effect, try pushing down on the end of a flexible ruler. For compressive strength, a component must have sufficient thickness to resist this buckling. A thickness of about 1/20 to 1/50 of its length is ideal. (i.e. say a post was 2 m (80 in) long, it should be 2 m × 1/20 or 100 mm (4 in) thick), i.e. have a 100 mm diameter. Timber columns should be thicker than steel columns. In terms of weight and strength, a steel pipe is very efficient because it has a large diameter without being heavy.

Bending

The third load type is caused by a sideways load on a beam such as a beam supporting a floor. In the floor, the beam is usually held at both ends. The same effect occurs, though, if the beam is firmly held at only one end so that it 'cantilevers' into space.

To observe bending in action, support a flexible ruler at each end and load it in the centre with a container of water. Now try clamping one end of the ruler to a table and measuring the amount of water needed to break it with the container at the end of the ruler. When clamped there is much more severe bending, and it will only take 0.25 as much load. This is similar to the case of a trellis post embedded in the gound or an unbraced cross-arm.

Note that as the ruler is loaded it bends. If we could accurately measure the distance around the outside and inside of that bend we would find the outer face to be slightly longer than the original length (i.e. under tension) and the inner face to be shorter (i.e. compressed). The factors influencing bending strength are as follows.

Material strength: The most common trellis materials are high tensile steel, mild steel, hardwood, softwood, and knotty softwood. Their relative strengths are approximately in the ratio 70,40,4,2,1. In other words, a solid mild steel bar would be 40 times stronger than a knotty timber of the same size. The knots are only a major factor when they are on the tension face of a beam because it is here they are pulled apart causing a crack. On the compression face they are squeezed and hence have little adverse effect on the strength of the beam.

Shape: For strength, a beam must be deep, and wide enough to resist buckling. 'C' section steel beams are particularly susceptible to buckling. Timber beams are susceptible to buckling if they are thinner than one third of their depth. A similar ratio applies to steel beams, and those with 'C' or 'Z' sections are particularly prone to buckling failures. To compare two beams of similar materials multiply the depth squared by the width. For example a '4 × 2' beam has a strength of 4 × 4 × 2 = 32 compared to a '3 by 2' with 3 × 3 × 2 = 18. In other words a '4 by 2' is nearly twice as strong.

Orientation: In the shape example, the depth was taken as the larger dimension. A '4 × 2' on its flat though has a strength of only 2 × 2 × 4 = 16, so correct orientation to the load is clearly most important.

Soil Effects

Soil classification

To minimize the danger of undersized anchors for end assemblies, it is worthwhile taking the trouble to check and classify soils across the proposed vineyard.

Clays: Clay soils can be identified by making a golf ball sized ball with the soil and a little water. The sample will feel smooth and putty-like with some stickiness.

Silts: Silts are also easily balled, but will not feel sticky. If the ball is squashed it will crack around the edges.

Sands: Pure sands will not be able to be made into a ball. Loamy sands can be, but will feel much grittier than the silts.

Soil strength

Strong soils: Strong soils will only show faint heel marks if walked on when wet. They generally contain a lot of clay. This binds the soil together rather like concrete.

Medium soils: Medium soils show a distinct imprint if walked on when wet. Most sands, gravels and gravel/clay mixtures containing predominantly gravel fit in this category.

Weak soils: When these soils are wet a thumb can easily be pushed into them. Often you can pinch a ball of them in half between the fingers. These soils are usually silts or soft clays.

In approximate terms, strong soils are four times stronger than weak soils and twice as strong as medium soils. This relates directly to their holding power for anchors.

Soil anchor selection

The two most common types of soil anchors are vertical posts driven into the ground, and steel plates attached to a shaft screwed into the ground (screw anchors).

Vertical post anchors: The capacity of the vertical post as an anchor can be increased by deeper embedment which increases the bearing area and hence soil resistance. The post diameter should also be larger, as there is more bending in the post from its greater length and increased load.

Typically a driven softwood post should be about ten times as long as its diameter if the post is pulled from no more than about 200 mm (8 in) above ground level. The main strength advantage of driven anchors is the fact that they are pushing against soils at several depths. Most soils are layered and each layer has a different strength. A driven post can pass through several layers and tends to average the soil strength of the various layers. A plate anchor relies largely on the strength of the soil immediately around the plate to restrict movement through the soil. This makes plates much more variable in strength than driven anchors.

Screw anchors: The strength of a screw anchor increases with the area of the screw plate, and also the depth at which the plate is located. If the soil is reasonably uniform with depth, increases in length beyond six times the plate diameter do not increase the anchor's capacity. For example, for maximum efficiency, a 150 mm (6 in) plate should have 900 mm (36 in) shaft. Well made screw anchors should be hot-dip galvanized to minimize corrosion, be securely welded plate to shaft and at the eye, and have a bevelled cutting edge on the plate. In hard soils, a heavier shaft is always an advantage in that it allows more torque (turning force) to be used on the anchor to screw it in. Screw anchors should be inserted on such an angle so that the rod is in the line of pull of the anchor ties. This will eliminate the movement of the rod being pulled into line once the tension is applied.

Which anchor?

To choose between anchors, use the sizes on the tie-back diagrams in the following sections as a guide to size for equivalent strength. In addition to purchase cost, consider installation costs. Very strong soils are difficult to drive posts into. Very stony soils are not suitable for screw anchors as the stones can be caught on the cutting edge of the plate.

A soil anchor. (Photo M.R.)

Failed end assemblies due to inadequate embedment. England. (Photo R.S.)

Testing end assemblies to determine ability to withstand load. Hamilton, New Zealand. (Photo M.R.)

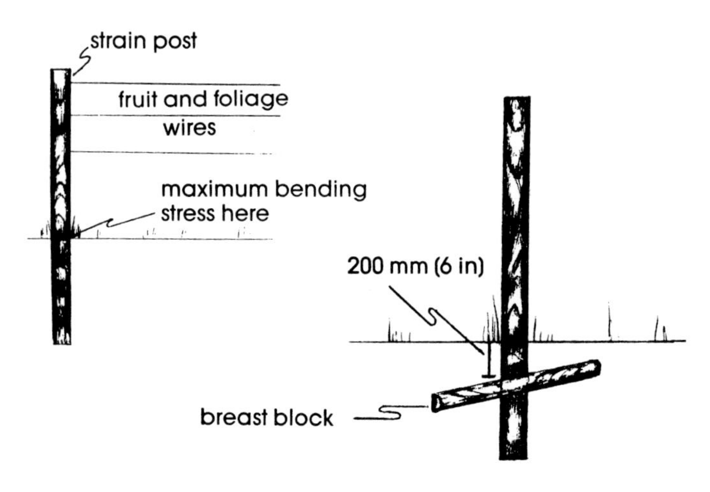

Free standing end assembly with optional incorporation of a breast block. Typical components are: strain post 200–250 mm (8–10 in) diameter, 1.2–1.5 m (4–5 ft) in ground; breast block 200 mm (8 in) diameter half round, 1.2 m (4 ft) long.

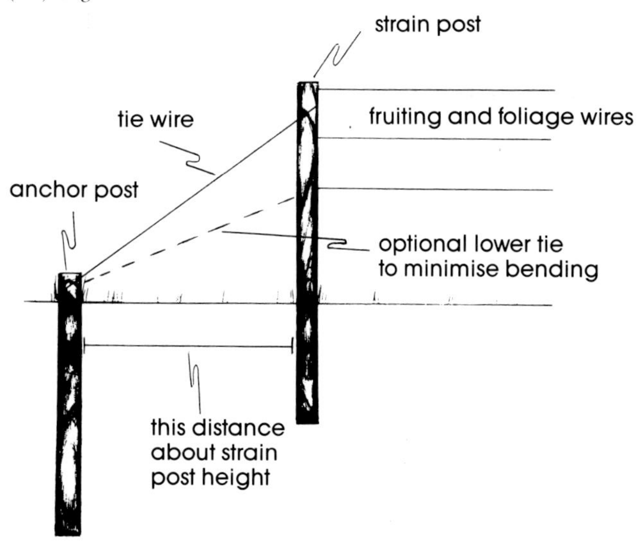

Tie back assembly. Typical components are: strain post 100–125 mm (4–5 in) diameter, 0.6 m (2 ft) in ground; tie 4 strands 2.5 mm (12 ½ gauge) high tensile wire; anchor post 175 mm (7 in) diameter, 1.8 m (6 ft) in ground for weak soils, or 150 mm (6 in) diameter, 1.5 m (5 ft) in ground for strong soils; optional screw anchors are 250 mm (10 in) diameter for weak soils, 200 mm (8 in) for medium soils and 150 mm (6 in) for strong soils.

A neatly-constructed tie-back system in South Australia. Note that the tie wires have been twitched, which can cause problems where overdone. Generally, this practice is not recommended. Also, the anchor posts are unnecessarily long which is wasteful of timber. (Photo R.S.)

Choosing an End Assembly–I Free Standing Posts and Tie Backs

The failure of an end assembly will affect the whole trellis. For this reason it is worthwhile taking particular pains over the selection of materials and construction. All assemblies will move a small amount when the trellis wires are tightened, but if they move more than 25–30 mm (1–2 in) the wires will go slack. When this happens, the assembly is regarded as having 'failed'. They fail by pushing through the ground, by jacking out of the ground or, if the post is too light, by breaking. Their manner of failure depends on the soil and the relative sizes of their various components.

In New Zealand and overseas tests have been done on end assemblies to determine which is the best. The variability of soil however makes comparisons very difficult. As a consequence it is very hard to make blanket recommendations.

Instead, basic designs are presented which are suitable for a typical vertical trellis system. With these basic designs are the usual range of sizes suitable for the different soil types. A good guide to optimum post size will also come from observing existing assemblies in similar soils. The posts illustrated are natural round softwood. Peeler cores should be about 20% larger, hardwoods could be about 20% smaller.

There are four major types of end assembly. They are free standing post, tie-back assembly, diagonal stay assembly and horizontal stay assembly.

Free standing posts

This assembly is simply a post driven into the ground. It relies on distance in the ground to avoid being pulled over, and diameter to avoid breaking. It is obviously easy to install, but is usually expensive because of its size and it tends to move under load. The post is often angled back so that the movement is less obvious. One way of minimizing the movements is to place a 'breast block' in front of it at subsoil level. This horizontal log spreads the soil pressure over a larger area. Apart from additional cost, the main disadvantage of the breastblock is that it concentrates the bending load at the breastblock height. This will quickly show up any weak posts!

Tie-back assembly

The tie-back assembly is the most efficient assembly in terms of strength and cost. It is also easy to build. Tie-back assemblies are often the most convenient to use in retrofit situations provided headland width is sufficient. Its strength comes primarily from the quality of the soil anchor. The strain post is little more than a prop although it does need sufficient diameter to attach wires without strangling the post, and enough strength to resist bending in the middle. By adding a lower tie wire, the bending stress in the middle of the post is dramatically reduced.

The remaining key component in the assembly is the wire tie from post to anchor. As a general rule, the tie should be as strong as the sum of all the wires tied to the post (count foliage wires as half). In other words, if you had two fruiting wires and four foliage wires, use at least four tie wires (two loops). This rule applies when the distance between the base of the post and the anchor is about the height of the post. Closer distances dramatically increase tie tensions.

Tie-backs commonly fail through poorly tied wires, undersized soil anchors and prop posts that are too slender.

Choosing an End Assembly–II Diagonal and Horizontal Stay Types

The main advantage of the diagonal and horizontal stay assemblies over the tie back assemblies is that they provide a vertical end to the row and thus maximize the productive space in your row. They are, however, more expensive as they use more materials to achieve the same strength. Construction time is also longer.

The stay assemblies work because they create a frame which is then required to pivot in the ground if it is to fail. Therefore to work well they need soil friction to hold the back post down, and to be rigid.

Diagonal stay

The diagonal stay pivots about the stay block, and is held at the back by the foot, which is dug in by hand or nailed to the base of the post. This assembly is not normally used with driven posts. The size of the stay block is also critical as enough bearing area is needed to ensure the block is not pushed through the ground. Undersized foots and stay blocks are the most common problems with these assemblies.

Horizontal stay

The efficiency of the horizontal stay assembly lies with the top rail and diagonal tie. They combine to form a rigid triangle which keeps the strain post vertical in the ground as shown in the diagram. Because the back post is kept vertical, this means that the load is spread more evenly over the soil. Since all the load is effectively transferred to the soil through the strain post, this should be the largest post. The strain post must be sufficiently deep in the ground to resist the uplift. This depth generates friction between the post and the ground and thus is the key to the operation of the assembly.

The uplift force is dependent on the length of the rail—longer rails mean less uplift. The inside post is only needed to hold up the front corner of the triangle formed by the rail and the diagonal tie. In fact, the front post is pushed into the ground when the load comes on.

Horizontal stay assemblies can fail in a variety of ways, but the most common are leaning forward caused by insufficiently strong diagonal wires, and uplift of the back post when it is not deep enough in the ground.

End assemblies for divided canopies

For trellises with horizontally divided canopies such as the 'lyre' there is a further complication, namely, how to reduce their width to a single point (i.e. the anchor).

As you can see from the drawing, the cross-arm in A is holding the wires up. This is bad in that a lot of bending load is put onto the arm. It is also inconvenient in that the wires close off a triangle. This then restricts the access of say a harvester or mower which you may want to operate close to the trellis posts from the end of the row.

Instead, use either the tie back as shown (B), or a horizontal stay with spreader bars on the front post. Another alternative, which eliminates cross-arms, is to drive two anchor posts at the end and bring the wires down to two anchors. This will best suit if the intermediate frames are also a two post system as the same shape is then duplicated into the end frame, thus maximizing machinery access options.

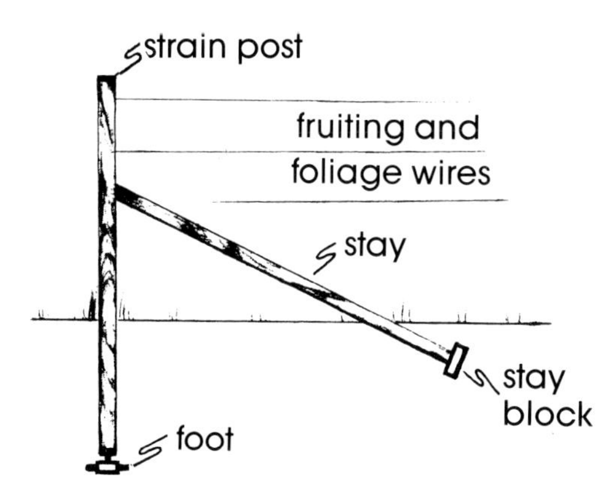

Diagonal stay end assembly. Typical components are: Strain post 150–175 mm (6–7 in) diameter, 0.9 m (3 ft) in ground; Foot 100 × 50 mm (4 × 2 in) block, 0.25 m (10 in) long; Stay 100 mm (4 in) diameter, 2.4 m (8 ft) long; Stay block 200 mm (8 in) face width half round, 1.2 m (4 ft) long.

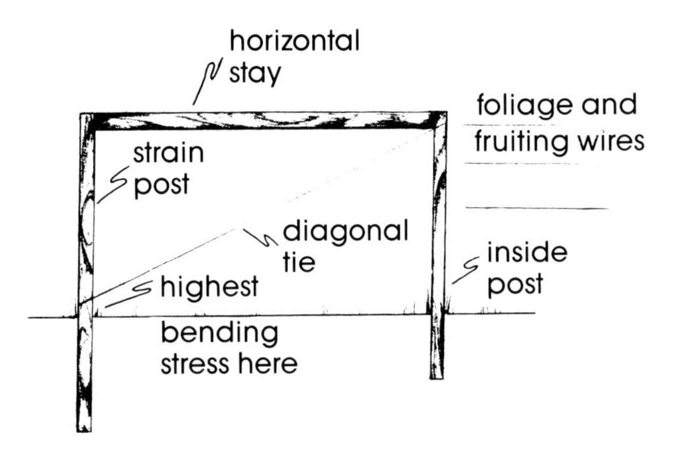

Horizontal stay end assembly. Typical components are: Strain post 150 mm (6 in) diameter, 0.9–1.2 m (3–4 ft) in ground; Horizontal stay 100 mm (4 in) diameter, 2.4 m (8 ft) long; Diagonal tie 4 strands 2.5 mm (12.5 gauge) high tensile wire.

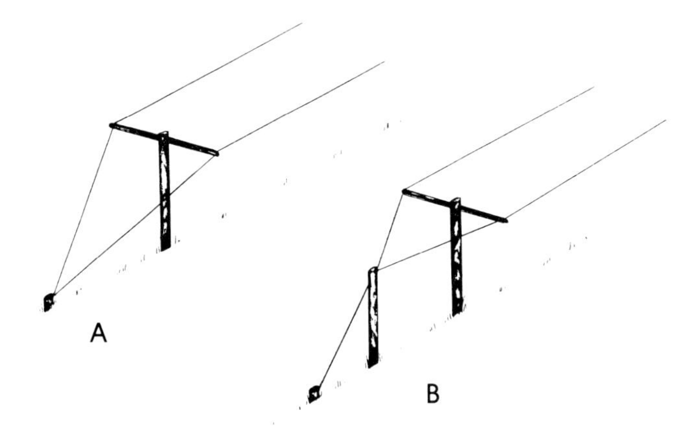

Incorrect (A) and correct (B) end assembly for horizontally divided canopy.

Wire Tensions and Sags in a Vineyard
Wire size: 2.5 mm high tensile steel (12.5 gauge)

	6 m (20 ft) post spacing		8 m (26 ft) post spacing	
	Heavy load*	**Light load**	**Heavy load**	**Light load**
Wire tension	2 500 N (550 lbf)	1 500 N (330 lbf)	3 000 N (660 lbf)	2 000 N (440 lbf)
Sag	180 mm (7 in)	130 mm (5 in)	250 mm (10 in)	200 mm (8 in)

* Heavy load 10 kg/m, 7 lb/ft
Light load 5 kg/m, 3.5 lb/ft

Typical Wire Strength Characters

Size High tensile wire	Maximum safe load	Break load
1.6 mm (16 gauge)	1 500 N (330 lbf)	2 200 N (480 lbf)
2.0 mm (14 gauge)	2 700 N (600 lbf)	4 000 N (880 lbf)
2.5 mm (12.5 gauge)	4 000 N (880 lbf)	6 000 N (1320 lbf)
3.15 mm (10 gauge)	5 700 N (1250 lbf)	8 400 N (1850 lbf)
Mild steel wire		
2.5 mm (12.5 gauge)	1 500 N (330 lbf)	2 100 N (460 lbf)
4.0 mm (8 gauge)	4 000 N (880 lbf)	6 000 N (1 320 lbf)

A neat and efficient tie off on an end assembly. (Photo M.R.)

Broken wire due to poorly tied knot. (Photo M.R.)

Wires for Trellises

On a strength for cost basis, by far the best choice for trellising wires is galvanized high tensile steel wire. Steel is intrinsically strong, relatively economical and readily connected, so it is an obvious choice. High tensile steel is preferred because steel pieces are closely related to weight, so it makes sense to get as much tensile strength per unit weight of steel as possible. To protect the steel from rusting, galvanizing is best because trellising wires have many places where rust could occur.

To make good use of these wires, however, particular care needs to be taken with their installation. This is because the wires are selected to be near their maximum capacity for optimum economy. As a result there is little margin for error. A badly chosen (but well tied) knot for example can reduce the effective wire strength to less than 50%. Poorly tied knots therefore need no discussion! Poor handling can also kink the wire or excessively damage its rust protection with similar consequences. Handling techniques are discussed under construction skills later.

When wire problems occur, they invariably can be classified into two categories—excessive sag or breaks.

Wire sag

When a wire is loaded downwards it sags under the load and the tension in the wire increases above the initial construction tension. If you wish to have less sag, the tension in the wire must be increased, the load reduced or the span reduced. To understand this relationship thread a short length of string through a weight such as several nuts. Now hold each end of the string and feel the tension increase as you try to reduce the sag by pulling the ends apart. Try the same process with a reduced span.

The table gives a range of tensions and sags for a heavily loaded (10 kg/m, 7 lb/ft) and lightly loaded (5 kg/m, 3.5 lb/ft) fruit wire for two common post spacings. By increasing the tension of the wire when it is first installed above the normal 1000 N (220 lbf) it is possible to decrease the sag. This is acceptable for the 6 m (20 ft) post spacing unless a stronger wire is used to handle the considerable extra tension that is generated when that taut wire is loaded with fruit. Note that if this happens the end assemblies will also need strengthening.

Sag is a natural phenomenon which must be accepted in the vineyard but should be within the limits of the table. If wires have sagged further than this, but all other variables are the same (i.e. wire size, load, span) your problem will be end assemblies which have moved, or insufficient tension at the start of the season.

Broken wires

If your vineyard is having problems with broken wires, there are a number of possible causes. Overloading can come from excessive tensions at the start of the season, or a poor choice of wire for the span or load. As a general rule 2.5 mm (12.5 gauge) high tensile steel wires are best for the fruiting wires. This wire has a breaking load of around 6 000 N (1 320 lbf) and a normal 'safe' load of 4 000 N (880 lbf). Foliage wires can be lighter, and 2.0 mm (14 gauge) which is about 40% weaker (and cheaper) is a good choice. Note that wire strengths and quality for a given size can vary with manufacturer.

Poor knots are a very common cause of breakages. Knots can be badly chosen or badly tied. For high tensile wires the two knots shown on the construction skills section are best. Their strength is about 80% of the break load of a plain (untied) wire. This is an acceptable reduction. Some other knots have strengths as little as 40% of the strength of a plain wire. Badly tied knots occur when the wire is bent sharply or is nicked with pliers. Both treatments create a concentration of stress at a sharply defined point which is where the wire will break.

Another cause for breakage is rust. The rust may have come from contact with an ungalvanized steel component or contact with a recently treated timber post. Rust can also occur if the wire has been poorly stored before it was used. The most common traps to avoid in storage are proximity to fertilizers or ground contact for a long period of time.

Intermediate Frames

The design of an intermediate frame is essentially that of creating supports for wires at predetermined points in space. These points and some basic frame designs have been discussed in the sections on trellis types. Trellises have been built from a wide range of materials including timber, steel, aluminium, PVC and concrete. In most cases, though, unless the material is surplus from some other project, timber and steel are the most economical options. Typical embedment depths of 75 to 100 mm diameter (3 to 4 in) wooden posts are 0.6 m (2 ft) for medium soils. See in the construction skills section for more details.

Steel frames

Because of the economics of mass production, steel suits kitset style frames with higher purchase price, but lower installation costs than timber frames. Steel posts may offer limited options for altering dimensions to suit your requirements. To maximize flexibility, it is prudent to choose a frame with more holes punched than you think you will need, as extra holes are difficult to drill—particularly if the steel is high tensile.

Secure fastening of components can be a strength or a weakness of a steel system. Welded joints are the strongest, but severely reduce options for varying the trellis in future. On the other hand, bolted connections are less secure, can be a moisture trap for rust, and are often difficult to install. ('U' bolt connections to pipes or stakes deserve a special 'frustration' award in this regard!). Some systems use specially developed clips. The best way to assess their application is to install about 10 with one hand (the other one holds the frame) and decide on fatigue and frustration levels.

Steel stakes are light and easily driven into stony soils. This dramatically increases the speed of construction. Their main disadvantage is that they are usually not as strong or as wide as typical 90 mm (3.5 in) timber posts. The lack of strength shows up in particular with 'C' section stakes which are susceptible to buckling. The width is important to minimize movement through soft soils. Both of these disadvantages are critical for divided canopies such as the lyre. This is because of the likelihood of unbalanced loads through the cross-arms which can cause a trellis to fall over. Ironically, the solution to this problem in fact shows steel stakes in their best light. If the divided canopies are built out of two stakes, the stakes can be connected to and support each other and provide support for the wires with a minimal requirement for extra connections.

The other situation where strength is important is for trellises which follow contours on hillsides. In this case, the curve of the row puts a sideways loading on the post which normally necessitates a larger diameter timber post. A cross-braced steel frame can achieve similar strength. The drawings illustrate some options for negotiating such corners.

The final check point for steel trellises is the facility to connect wires to the frames. Ensure that any touch-points between the frame and/or connection and the wire are galvanized to minimize rust of the wire. Also, check that the fruiting wire connection can take a downwards load of 80 kg (180 lb) (equivalent to 10 kg/m (7 lb/ft) times 8 m (26 ft) frame spacing). The foliage wire connection needs to be movable and resistant to possible harvester vibration but will require very little strength.

Timber frames

With timber frames the principal causes of problems are poor connections. Possible connectors to hold cross-arms to posts include bolts, self tapping screws, nails, nail plates, wire and staples, 'Z' nails, all with or without mortices in the post. All of them have their place. To test a connection system, simply hang 80 kg (180 lb) weights at each fruit wire position and vibrate the post as with a harvester. If it passes this test, then remove the weights from one side and repeat the test.

For those wanting recipe solutions, try an 8 mm (5/16 in) bolt for fruit wire cross-arms, two 75 mm (3 in) nails for each end of the braces and three 100 mm (4 in) nails for foliage wire cross-arms. It is also important to watch for splits in timber connections. These occur when bolts and/or nails are too close together, or too close to the edge. As a general rule, ensure bolts or nails are centred at least 15 diameters from ends and 5 diameters from edges of the timber. In other words an 8 mm (0.3 in) bolt holding a rigid cross-arm should be at least 120 mm (5 in) from the top of the post and 40 mm (1.6 in) from the edge of the arm (i.e. on a 75 mm (3 in) cross-arm, put it in the middle).

The same rule applies to staples and nails, but you have to consider the grain in the timber. The minimum across-grain spacing between nails or staples is 5 and the along-grain spacing is 15 times the diameter. For example a 75 mm (3 in) nail has a diameter of 3.15 mm (0.13 in), so the minimum spacing is 16 mm (0.6 in) across and 50 mm (2 in) along the grain.

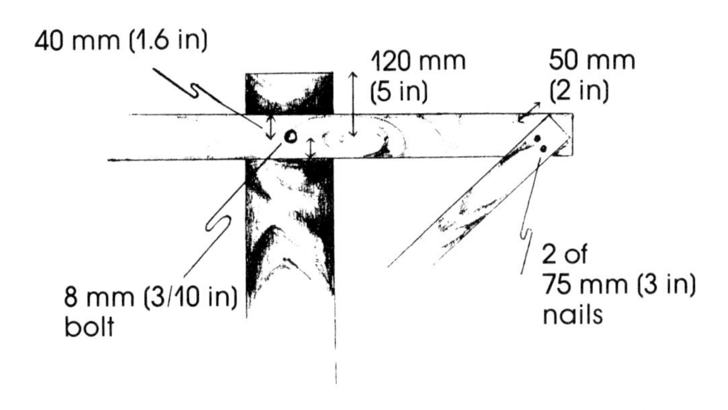

A schematic cross-arm and brace indicating nail and bolt placement criteria.

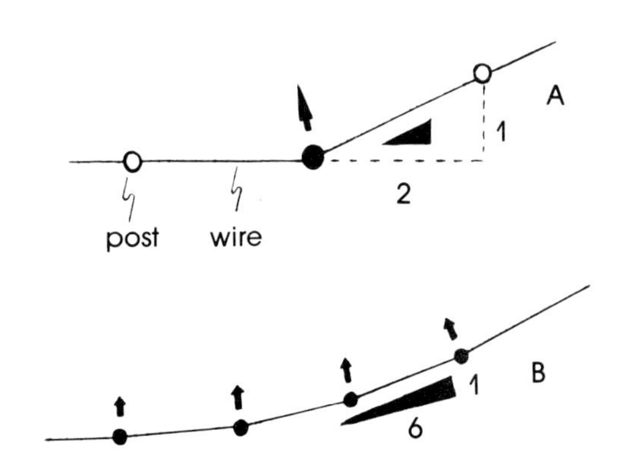

A & B are plan views of two methods of following a contour. The sharp turn in A produces a side load on the back post which in this case is the same as an end assembly load! (So an end assembly should be built there.) Alternatively, Plan B uses more posts, but the load on each post is reduced to sensible proportions capable of being held by a 100 mm (4 in) timber post in most soils. The triangles show how to measure the sharpness of the turns.

A failed end assembly. California. (Photo R.S.)

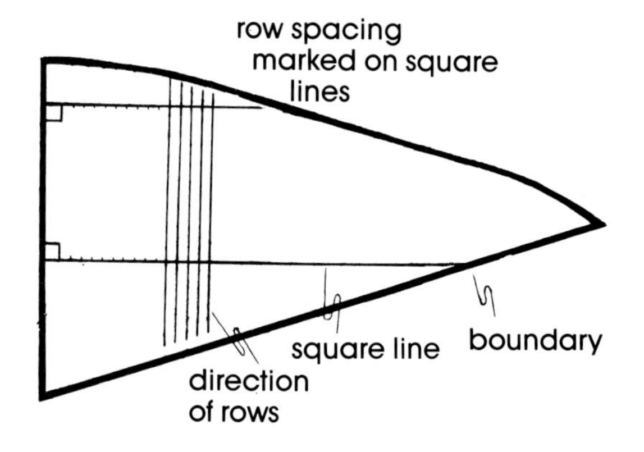

Marking out a vineyard with irregular boundaries by establishing two parallel lines normal to one boundary.

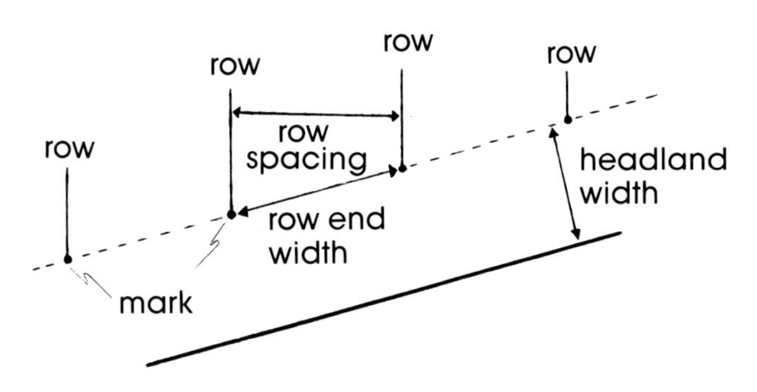

Careful layout of headlands and row ends, using a string or wire marked with row end width.

This logo is a trademark which may be used by fencers who have completed the 'Master Fencer' course conducted by Wiremakers Ltd, New Zealand.

Trellis Construction Using Contractors

When building a fence for animals, the load on the wires when it is built is invariably the heaviest load it will ever have. For this reason, if you can build a stock fence and it survives its first day, it will probably last quite well. Unfortunately, the same is not true for grapevine trellises as their peak loads occur when the vine is fully developed several years later, so failures can occur at the worst possible time for the owner. A key consideration toward achieving a quality trellis is to have quality construction. Attention to detail and awareness of potential errors can go a long way towards a frustration-free trellis.

The task may seem a mundane one which anyone could do—but the opposite is true. In New Zealand, a national training course called the 'Master Fencer Scheme' has been set up to provide standards for fencers. The level of understanding and quality of workmanship of fencers who have attended the course has proved to be a great boon to growers who have contracted them to work on their properties.

Trellis design

It is important to resolve who is to design the trellis so that it is clear who is responsible for any subsequent failures. For this you could ask the contractor, do it yourself, employ a specialist engineer experienced in trellis work or buy a proprietary system complete with manufacturer's guarantee and detailed installation drawings. For all but the smallest jobs, the latter two are generally most worthwhile to minimze cost and downstream troubles.

Material supply

Who is going to supply the various components, and when, are questions with many variations of the 'what if' variety. For example, what if the supplier is late, the materials are of poor quality, the size of the job is reduced or increased? Resolve such questions before the contract is signed.

Construction scope

If you are planning to do some of the work yourself using limited equipment and semi-skilled labour, then using a fencing contractor to do key parts of the structure can be worthwhile. Consider layout and post driving, anchor installation, and wiring up end assemblies as possible tasks for the contractor. Alternatively, you may wish the contractor to finish everything to the smallest detail. In both cases detailed drawings or a completed row to be used as an example will minimize misunderstandings.

Layout

Given a plan, a contractor can lay out a vineyard unaided. This can be a risky and frustrating process for both you and the contractor unless it is *very* straightforward. It is much better to organize to be on site until the boundaries are established, critical accessways defined and corner posts are driven. Needless to say—check and re-check boundaries against legal surveys! It is also advisable to check headlands and row spacings before too many posts are driven as distances often seem much closer than they may have appeared on the plan.

Workmanship

Be aware that anyone who has strained a wire or dug in a post in his backyard may bestow himself with the title 'fencer'. Many will have little or no experience with horticultural trellises and all the subtleties they embody. Asking to see their previous work is perhaps one solution; alternatively ask for references.

Construction Skills

Layouts

An accurately laid out vineyard makes every subsequent job including planting and all mechanical operations much easier.

The first step is to find a corner to work from. Check it for square on an aerial photograph or using a surveyor's level or theodolite. If such a corner does not exist (such as in triangular blocks of land) run two square lines away from the edge with the longest row as shown in the diagram on the previous page. Sighting poles can then be put along the square lines to sight in the ends of each row. Alternatively, use trigonometry to calculate the width of the rows on the angle. In either case, the row width is marked on a piece of string or wire which is pulled tight along the square lines or end of the vineyard.

Mark the ground at measured spots along the string with a fluorescent spray paint. Use the same string or wire at the other end, and at any ridges or changes in direction down the row. The row spacing is then well defined all over the vineyard. The next step is to use the same method down the rows with a new piece of string or a tensioned wire. If you are using a tensioned wire, consider installing the anchor posts and running a wire along the ground before the rest of the posts are driven.

Embedment depths

The optimum embedment depth for intermediate posts is governed by the loads they must support, and resistance offered by the soil. Re-read the section on soil effects to assess the average strength of your soil in the region. Wooden posts of 75–100 mm (3–4 in) are typically embedded about 600 mm (24 in) for medium soils. If 600 mm (24 in) was a reasonable depth for medium soils then weak soils would need 850 mm (34 in) and strong soils 460 mm (18 in) for the same resistance to overturning. Add an additional depth of 100–150 mm (4–6 in) if any of the following conditions applies: narrow posts (i.e. steel stakes), wide support structures (e.g. GDC), windy site or exposed boundary, very wet site (or sprinklers near posts), or very soft soils or deep topsoils. You can deduct 100–150 mm (4–6 in) for twin posts in a divided canopy, for post spacings less than 6 m (20 ft) or posts less than 1.5 m (60 in) height.

Wiring end assemblies

The skill here is to install the wire so that it is tight and hence the post is at the required angle when under load. The wires should be tied so they will not slip down the post. This problem is usually caused by a steep tie-off angle or by poor stapling. For maximum strength the wire should only be tied to the post (i.e. no knots in the wires from post to anchor). The simplest end assembly to build is the tie-back.

To wire a tie-back assembly, the first step is to put a holding point on the anchor and the first post (e.g. a staple). Then thread as many loops of wire loosely between anchor and post as are needed, leaving the two ends with about 500 mm (20 in) spare hanging down from the post. Standing on the anchor side of the post, pull on these two ends so that the top of the post comes towards you. Then drive the back staple in to hold the post in position while you tie the wires off. They can be tied together on the side of the post facing you. Any knot will do, but try the very neat knot shown on the diagram. When the fruiting wires are installed and tensioned, the post will be pulled back to its original position and the tie-back wires tightened. If you are using a driven timber anchor, it is easier to do the tie-off on the anchor.

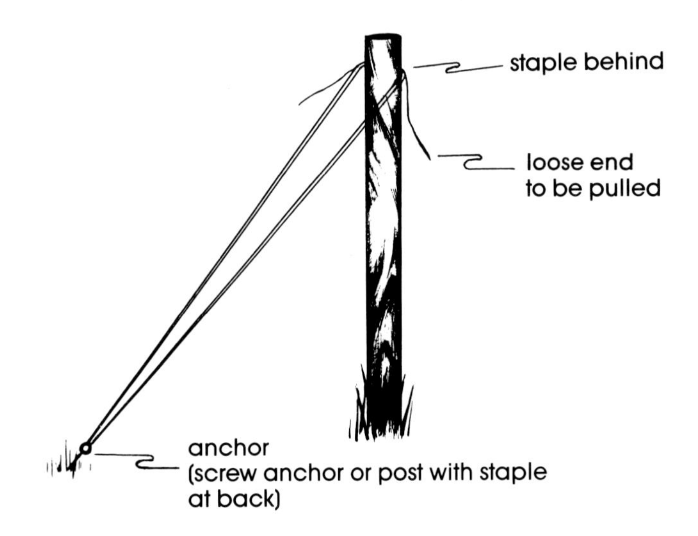

To wire a tie-back assembly. Loop sufficient wire between anchor and post, leaving two loose ends (500 mm, 20 in).

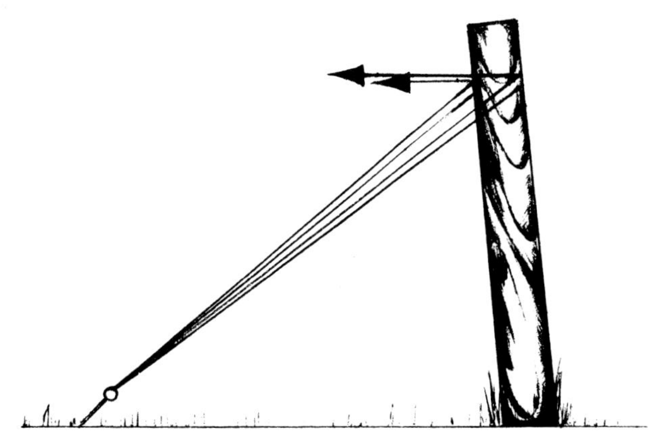

Pull the wire tight moving the post towards you.

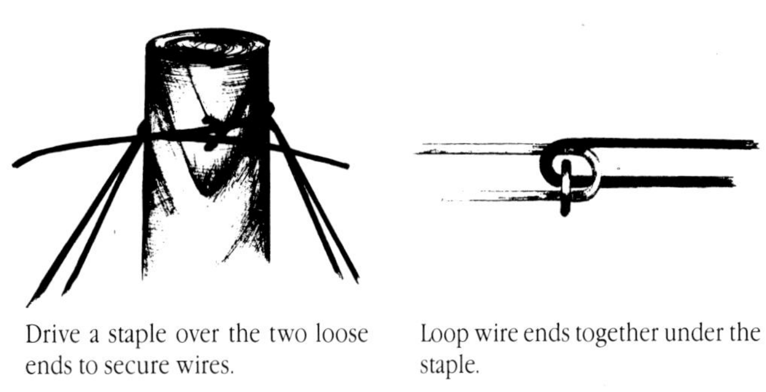

Drive a staple over the two loose ends to secure wires.

Loop wire ends together under the staple.

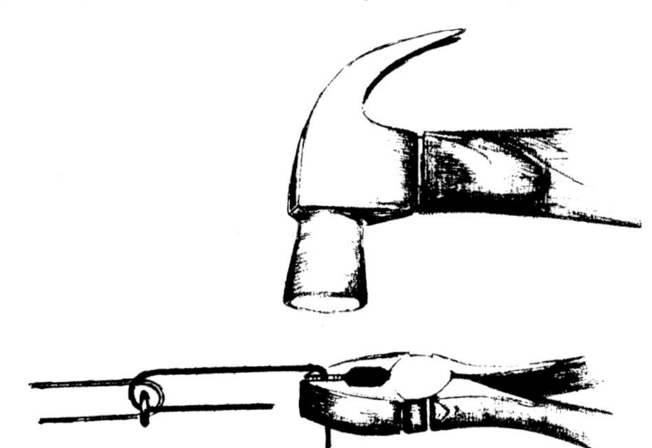

Cut the loose end and bend into a 25 mm (1 in) L. This can then be driven into the post by holding it at the start with the pliers.

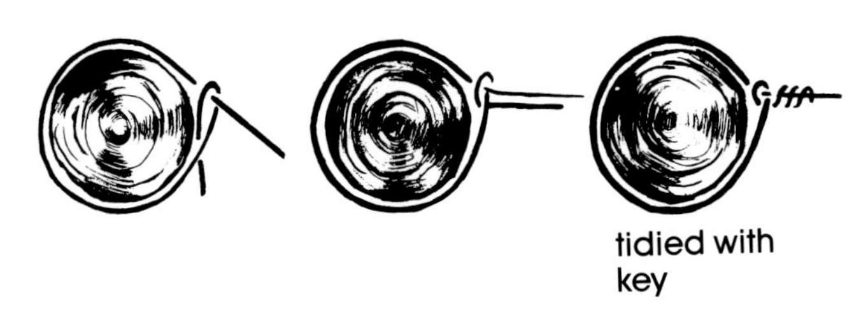

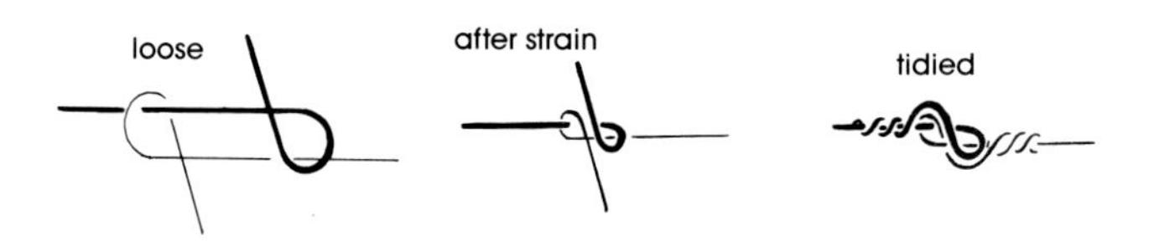

Post tie off and wire joining knots.

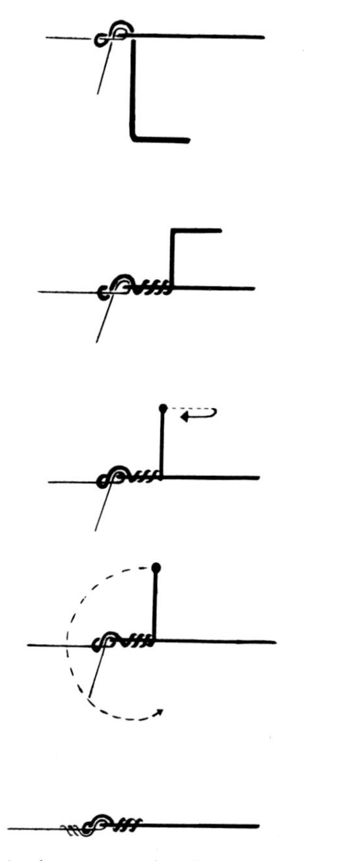

Tidying up the joining knot. Form a handle with the end 100 mm (4 in) to wind on several tight coils, then twist the handle to break it off cleanly at the knot.

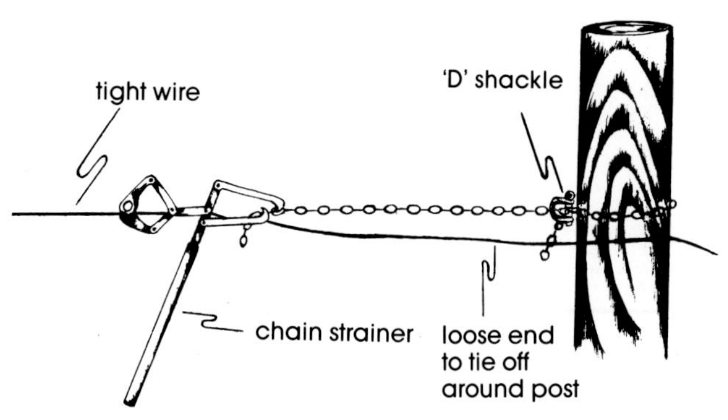

Tensioning a wire to a post.

Anchors

Correct anchor selection is a key feature of good design. A good operator should, however, be able to spot and make allowance for variations in soil type. Straightness and cleanness of installation can also enhance the holding capacity of the anchor as a minimum of soil is disturbed.

Posts

The comments for anchors also apply to posts. Note that driven posts are invariably firmer than dug in posts. In stony ground consider using steel stakes as they can more easily find the gaps between the stones than timber posts. Another alternative in hard or stony soils is to drive a steel spike about ¾ as thick as the timber post first—make it slightly tapered so that it is easily removed—then drive the post in the same hole. Cutting a point on the end of timber poles can also help if driving is difficult.

Installing cross-arms

For timber cross-arms the timber should be oriented into its strongest position (i.e. edges with knots facing down) and firmly fastened. If mortices are to be cut into the posts, use a truck or platform to stand on so that the chainsaw is never above waist height.

Fruiting wires

These must be firmly tensioned to about 1 000 N (220 lbf) and neatly tied. To check the tension, use a tension meter, or try hanging a 4 kg (10 lb) weight from the centre of a 6 m (20 ft) bay—it should deflect the wire about 75 mm (3 in). The reason for testing tension is that slack wires will eventually become more slack and need retightening, and over-tight wires will lead to over-loaded wires and end assembles. This can result in costly breakages.

The wires should also be tied off neatly. Snag ends might damage a workers eye and poorly tied knots might cause a wire breakage. 'Allowable' knots for high tensile wires are shown in the diagram. Alternatively, use a proprietary wire fastener or tensioner.

There are a wide variety of permanent wire tensioners available. Their main benefit is that they allow easy re-tensioning of the wire. Before deciding on the expense of such a purchase, note that a correctly chosen and tensioned wire tied to secure end assemblies will not need retensioning. If you do have wires that become slack, buy tensioners of the type than can be put on without cutting the wire. They can then be used on only those few wires that need them.

The other consideration in choosing permanent tensioners is their ease and safety of operation. Look for models that have winders which can be operated with one hand (the other hand is then free for emergencies); clips and pins that are attached to the body of the tensioner (so they don't get lost or fall out of your hand as you are about to put them in position); and a bare minimum of sharp edges (which can cause worker discomfort).

Use chain wire strainers to tighten the fruit wires. Take the chain strainer and chain with D schackle attached to make a throttle-type strop. Put the strop around the post and strain up the wire to the correct tension. Give the strainer two extra clicks then tie off as shown—the extra clicks will compensate for the loss of tension in tying off.

Foliage wires

Foliage wires are commonly put in loose with permanent wire tensioners on them to adjust tensions as the wire is moved. A neat alternative is to tie the wire off with a tension of about 500–1 000 N (110–220 lbf) when it is in its central position on the post or structure. On all but the shortest rows, the stretchy nature of the wire will allow it to be pulled up, down or out, to other positions with minimal increases in tension.

A Selection of Wiring Equipment

Fencing pliers
Fencing pliers can be used for cutting, holding or bending wire and pulling out staples. Although they have a hammer face, using them as a hammer may make them seize. Using them as a staple pick by hitting them on the hammer face with another hammer may break the pliers through the pivot.

Claw hammer
For driving staples (not illustrated).

Side cutter or small bolt cutter
These have hardened jaws which make them particularly suitable for cutting high tensile wire. (not illustrated)

Chain strainer
Although a trellis can be erected quite easily with only one chain strainer, it is more efficient to have two or three. Some fencers prefer to use as many as there are wires on the trellis. The hooks on the chain strainers are spring loaded. The springs tend to get damaged, which often stops them from working properly. The chain strainer will work satisfactorily with the springs removed. When oiling strainer jaw hinges, be sure to wipe any oil off the parts of the jaw which grip the wire. As an addition to the basic unit, some strainers have a built-in tension gauge to help measure the tension on the wire. This can be used instead of a wire tension meter. You may also care to make a strop for straining to posts by removing the diamond jaws from the chain and attaching a 'D' shackle instead.

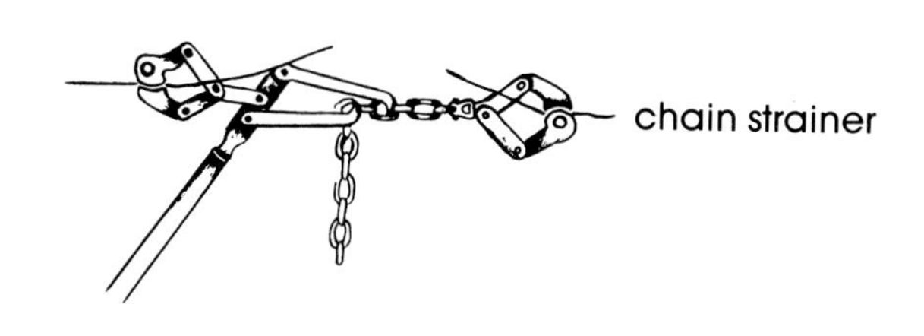

Wire tension meter
Although commercial models are available, this very important piece of fencing equipment is both simple and cheap to make. All you need is a piece of board 1 100 × 100 × 25 mm (44 × 4 × 1 in), two nails and a spring balance reading up to 10 kg (20 lb). The tension is measured by pulling the spring balance until the wire has deflected 12 mm (0.5 in). The tension in Newtons then equals the reading on the balance in kg multiplied by 200, or the tension in lbf equals the reading in lb multiplied by 20. Therefore, for a wire to be tensioned to say 1 000 N the reading should be 5.0 kg.

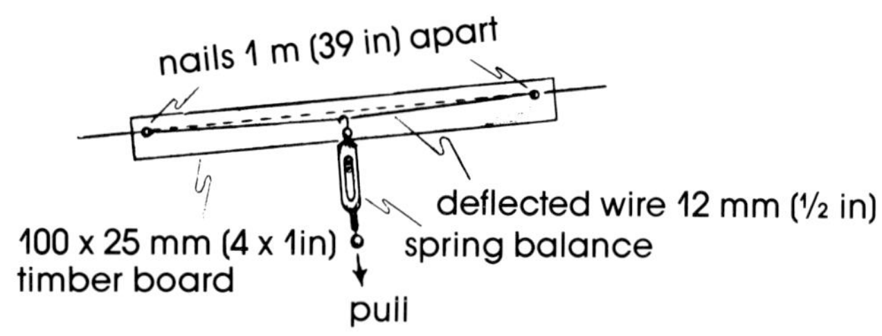

Wire tension meter constructed out of timber and a spring balance.

Wire keys
Wire keys are a tool for 'tidying up' a fence by wrapping the snag ends of wire round the line wire. The same job can be as neatly by hand as shown in the diagram on the previous page.

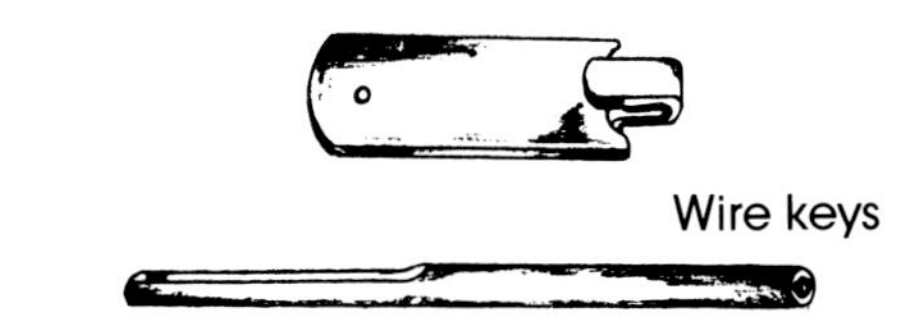

Wire keys for wrapping wire.

Spinning Jenny
Virtues to look for when buying (or making) a spinning Jenny include arms at least 600 mm (24 in) long (measured from the centre), with adjustments to take coils of different diameters. If you are running a lot of wire, a Jenny (or set of Jennies) which can be mounted on a trailer is very convenient.

When putting a coil of wire on a Jenny, find the free end (usually the one with the manufacturer's tag on it), keep it on top and mark it by bending a kink in it whenever wire is out off the coil. Do not remove the strapping wires holding the coil together until the coil is on the Jenny. Replace two of these wires before taking the coil off again to avoid subsequent tangles.

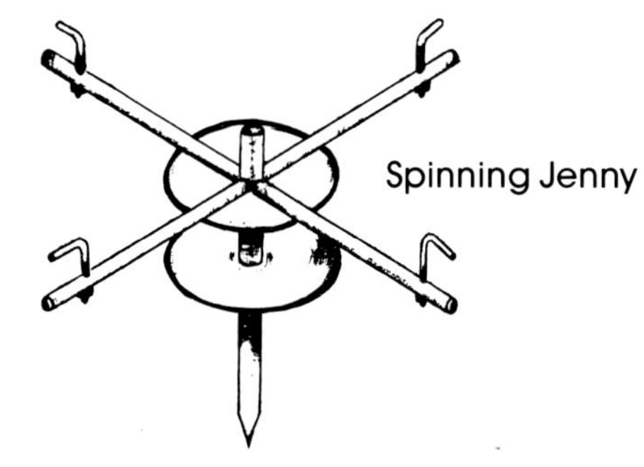

Spinning Jenny for laying out wire.

A well constructed end assembly. Hawkes Bay, New Zealand. (Photo R.S.)

An expensive trellis retrofit with two cross-arms on each single vine stake. California. (Photo R.S.)

A well constructed trellis system can also be used to support wind protection cloth in some circumstances. South Australia. (Photo R.S.)

Checklist for Quality Trellis

The notes from the preceding pages have indicated how a trellis takes the load, and factors which might influence its strength and durability. Clearly some qualified advice or at least a detailed look at working examples of a trellis similar to the one you wish to build would be most beneficial. To help you assess these examples for strength and durability, run the following check list over them to see if they will meet your standards of quality, strength, durability and cost.

Fruiting wires: are they firm/tight?
If yes, check that they have withstood a crop load as it is difficult to assess the long term life of an untested trellis.

Have the end assemblies moved?
Look along the back of them to see if they are still in a straight line with no gaps behind the posts in the ground, no broken tie wires and that there are no bent posts or arms. A strong and reliable end assembly should not have moved more than 25 mm (1 in) at the top of the post, when under full load.

Are the intermediate posts and frames too weak?
These will show as broken members, twisted joints or splits at staples. The twisted joints may still be holding, but will they interfere with future vineyard operations such as pruning and harvesting?

Will it rot or rust?
To check whether treated timber has been used, try prodding suspect timber with a screwdriver, particularly at joints. Are there moisture traps around the steel? Is there any ungalvanized steel, and are there any wear points for the wires?

Will it be difficult or unsafe to use?
Are there sharp points to poke eyes? Is there room for machinery to harvest/prune/maintain rows around frames? Are there loose or deeply sagging wires restricting machinery? Are wires in the correct positions needed to support fruit and foliage? Is there facility for moving wires as needed?

What is the real cost?
Assess capital cost plus installation plus allowance for repairs. Also make allowance for extra labour to work around awkward parts and potential lost time or crop if the trellis breaks.

Could the cost be too high compared to the benefits?
Good quality trellises need not be expensive. Equally, the benefits of an upgraded trellis are very clear. See a consultant for help in analysing the benefits in your situation.

Scott Henry trained vines side by side with VSP, showing larger canopy surface area and improved fruit exposure. Manutuke, New Zealand. (Photo B.W.)

Downward trained spurs as used for systems like GDC, Scott Henry and RT2TB. Manutuke, New Zealand. (Photo R.S.)

The interior of a grapevine canopy showing how petioles bend and twist so that leaves receive the most light. Cabernet Franc, Rukuhia, New Zealand. (Photo R.S.)

Vigorous young Chardonnay vines in their second growing season being trained to a metal RT2TB. These vines produced up to 12 m (40 ft) or cordon per vine by the end of the second growing season, due to continual vine training. Rukuhia, New Zealand. (Photo R.S.)

WINEGRAPE CANOPY IDEOTYPE

THIS SHEET CAN BE COPIED FOR USE IN YOUR VINEYARD

Character	Optimal value	Your vineyards and date				
Row orientation	north-south					
Ratio of canopy height: alley width	~1:1					
Foliage wall inclination	About vertical					
Bunch zone location	Near canopy top					
Canopy surface area (SA)	About 21 000 m^2/h (92 000 ft^2/ac)					
Ratio of leaf area: canopy surface area (LA/SA)	< 1.5					
Shoot spacing	~ 15 shoots/m (4.6 shoots/ft)					
Shoot length	10-15 nodes, 0.6-0.9 m length (2-3 ft)					
Lateral development	5-8 lateral nodes per shoot					
Ratio of leaf area: fruit weight	~12 cm^2/g					
Ratio of yield: canopy surface	cool climate 1-1.5 kg/m^2 (0.2-0.3 lb/ft^2) hot sunny climate to 3 kg/m^2 (0.6 lb/ft^2)					
Ratio of yield: pruning weight	5-10					
Growing tip presence after veraison	nil					
Cane weight	20-40 g (0.7-1.4 oz)					
Internode length	60-80 mm (2.4-3.1 in)					
Pruning weight per length canopy	0.3-0.6 kg/m (0.2-0.4 lb/ft)					
Proportion of canopy gaps	20-40%					
Leaf layer number (LLN)	1.0-1.5					
Proportion exterior fruit	50-100%					
Proportion exterior leaves	80-100%					

Further Reading

This list includes key references from the scientific literature which supports the ideas presented in this booklet. Included as well are references cited in the text.

Carbonneau, A. (1985) Trellising and canopy management for cool climate viticulture. Proc. First Int. Symp. Cool Climate Viticulture and Enology, June 1984 Eugene, Ore. Oregon State University, 158-74.

Clingeleffer, P. (1989) Update: minimal pruning of cordon trained vines (MPCT). Aust. Grapegrower and Winemaker 304, 78-83.

Intrieri, C. (1987) Experiences on the effect of vine spacing and trellis training system on canopy microclimate, vine performance and grape quality. Acta Horticulturae 206, 69-88.

Kliewer, M. and R. Smart (1988) Canopy manipulation for optimizing vine microclimate, crop yield and composition of grapes. In: C. Wright (Ed.) Manipulation of Fruiting. Proc. 47th Easter School in Agric. Sci. Symp. April 1988, Univ. Nottingham. Butterworths, London, 275-91.

Koblet, W. (1988) Canopy management in Swiss vineyards. Proc. Second Int. Symp. Cool Climate Viticulture and Oenology, January 1988, Auckland, New Zealand. NZ Soc. for Vitic. and Oenol., 161-4.

Morrison, J. (1988) The effects of shading on the composition of Cabernet Sauvignon grape berries. Proc. Second Int. Symp. Cool Climate Viticulture and Oenology, January 1988, Auckland, New Zealand. NZ Soc. Vitic. and Oenol., 144-6.

Reynolds, A. and D. Wardle (1989) Impact of various canopy manipulation techniques on growth, yield, fruit composition, and wine quality of Gewurztraminer. Am. J. Enol. Vitic. 40, 121-9.

Shaulis, N., H. Amberg and D. Crowe (1966) Response of Concord grapes to light, exposure and Geneva Double Curtain training. Proc. Am. Soc. Hortic. Sci. 89, 268-80.

Smart, R. (1982) Vine manipulation to improve wine grape quality. Grape and Wine Centennial Symposium, June 1980, Davis, California. University of California, Davis, California, 362-75.

Smart, R. (1984) Canopy microclimate and effects on wine quality. Proc. Fifth Aust. Wine Industry Tech. Conf., November-December 1983, Perth, WA. Aust. Wine Res. Inst., 113-32.

Smart, R. (1985) Principles of grapevine canopy microclimate manipulation with implications for yield and quality. A review. Am. J. Enol. Vitic. 35, 230-9.

Smart, R. (1985) Some aspects of climate, canopy management, vine physiology and wine quality. Proc. First Inst. Symp. Cool Climate Viticulture and Oenology, June, 1984, Eugene, Oregon. Oregon State University Tech. Publication 7628, 1-19.

Smart R. (1987) The influence of light on composition and quality of grapes. Acta Horticulturae 206, 37-47.

Smart, R. (1988) Shoot spacing and canopy light microclimate. Am. J. Enol. Vitic. 39, 325-33.

Smart, R., J. Robinson, G. Due and C. Brien (1985) Canopy microclimate modification for the cultivar Shiraz. I. Definition of canopy microclimate. Vitis 24, 17-31.

Smart, R., J. Robinson, G. Due and C. Brien (1985) Canopy microclimate modification for the cultivar Shiraz. II. Effects on must and wine composition. Vitis 24, 119-28.

Smart, R. and S. Smith (1988) Canopy management: identifying the problems and practical solutions. Proc. Second Int. Symp. Cool Climate Viticultre and Oenology, January 1988, Auckland, New Zealand. NZ Soc. for Vitic. and Oenol., 109-15.

Smart, R., J. Dick, I. Gravett and B. Fisher (1990) Canopy management to improve grape yield and wine quality—principles and practices. S. Afr. J. Enol. Vitic. 11, 3-25.

Solari, C., O. Silvestroni, P. Giudici and C. Intrieri (1988) Influence of topping on juice composition of Sangiovese grapevines (*Vitis vinifera L.*). Proc. Second Int. Symp. Cool Climate Viticulture and Oenology, December 1988, Auckland, New Zealand. NZ Soc. for Vitic. and Oenol. 147-51.

Winkler, A., J. Cook, W. Kliewer and L. Lider (1974) General Viticulture. University of California Press, Berkeley. 710 pp.

The Master Fencer Scheme

The Master Fencer Scheme has been developed by Wiremakers Ltd, NZ Agricultural Engineering Institute and NZ Ministry of Agriculture and Fisheries to be a link between good product and good workmanship.

For more information about the course, contact:

'Master Fencer Scheme'
Wiremakers Ltd
PO Box 22198
Auckland
New Zealand

This is meant to be GDC! Shoot positioning required! (Photo R.S.)

If you find yellow leaves when you remove the outside leaf layer, expect yield and quality problems. (Photo R.S.)

This 'V' trellis is far too narrow at the base to be considered a divided canopy. (Photo R.S.)

HELP! Send for a Canopy Management Consultant.